Alternate Realities

How Science Shapes Our Vision of the World

Alternate Realities

How Science Shapes Our Vision of the World

JOEL DAVIS

PLENUM TRADE • NEW YORK AND LONDON

Library of Congress Cataloging-in-Publication Data

Davis, Joel, 1948-
Alternate realities : how science shapes our vision of the world / Joel Davis.
p. cm.
Includes bibliographical references and index.
ISBN 0-306-45629-X
1. Science--Philosophy. 2. Reality. 3. Vision. I. Title.
Q175.32.R42D39 1997
530'.01'9--dc21 97-26417
CIP

ISBN 0-306-45629-X

Plenum Press is a Division of Plenum Publishing Corporation
233 Spring Street, New York, N.Y. 10013-1578
http://www.plenum.com

10 9 8 7 6 5 4 3 2 1

Printed in the United States of America

For Judy

Acknowledgments

This book was made possible in large part by the generous help and encouragement of many people. For their assistance I thank Dr. David Kliger and Dr. Eugene Switkes, University of California, Santa Cruz; Dr. Jerry Nelson, W.M. Keck Observatory, Hawaii; Dr. Steve Vogt, University of California, Santa Cruz/Lick Observatory, California; and Dr. George Smoot, University of California, Berkeley. All of them graciously consented to be interviewed at length for this book. They also reviewed various portions of the manuscript in progress and offered invaluable suggestions, corrections, and overall encouragement. Thanks, also, to the many men and women who reviewed the finished manuscript and offered comments and corrections, especially Dr. Charles Bennett, NASA Goddard Space Flight Center, Greenbelt, Maryland. Any errors of fact herein are not of their doing.

I also thank the librarians at Eastern Washington University, Cheney, Washington; Gonzaga University, Spokane, Washington; at the public libraries in Bellingham, Olympia, Spokane, and Spokane County, Washington; at The Evergreen State College, Olympia, Washington; and at the University of Washington Libraries, Seattle, Washington. Libraries are our national and local treasures; support them and support our freedom to read.

Bill Cannon in the Office of Public Information at the University of Washington was invaluable over the years in helping me find answers to some of my questions, including on this project. Just as helpful on this book were Jay Aller and Pat Orr at the California Institute of Technology in Pasadena, California; Justin Harmon, Director of Communications/Publications, Princeton University; Julietta Gonzales of the University of Arizona in Tucson; Robert Irion and Jennifer McNulty of the University of California, Santa Cruz, Public Information Office; Laurel Phoenix, Lick Observatory; and especially Andy Perala at the W.M. Keck Observatory in Hawaii.

Special thanks go posthumously to my first wife, Marie Celestre. Through fifteen years of partnership she supported my career as a freelance writer and encouraged my large-scale foray into the world of nonfiction popular science books. Though we eventually divorced, we still remained in friendly contact with one another. She was pleased and interested in the subject of this book when I told her about it. I deeply regret that she died before it was completed, and never had a chance to read it.

I thank my agent, Joshua Bilmes, for his patience and firmness in shepherding me through this book, and who offered me much encouragement throughout its creation.

Many friends and relatives offered encouragement during the time I worked on this book and I thank them: my parents, Jerry and Toni Davis; my brothers Chris, Mickey, Tim, Peter, and Ed, and my sister Marie Wights and her husband Randy; and my stepdaughter Kirsten Nash and her husband Paul. Thanks also to John Celestre, Millie Celestre, Dr. John Cramer, John and Gail Dalmas, Tony and Ida Dolphin, Dr. Mickey Eisenberg, Elana Freeland, Dr. James Glass, Megan Lindholm, Vonda McIntyre, Laura Sandberg, Theresa Scott and Rob Whitlam, John and Lee Seaman, Sara Stamey, and Tom and Jan Westbrook.

Finally, and most importantly, I thank my wife, Judy. Her kind and continuous support has been invaluable during the writing of this book. I could not have completed it without her help.

Contents

1

In a Grain of Sand

To see a World in a Grain of Sand
And a Heaven in a Wild Flower,
Hold Infinity in the palm of your hand
And Eternity in an hour.

—William Blake
Auguries of Innocence

Poetry and science are closer than most people realize. Many poets and scientists already know this, of course. Most of the rest of us are still trapped in dismal stereotypes about both fields of human endeavor. The deep link between the two is *vision*.

The Inner Eye

By vision I do not mean merely physical vision, the sense of sight. That is only the beginning. Rather, I refer to the "inner vision" or "imaginative vision" that each of us possesses. It is based on physical sight only to the extent that sight is the dominant sense for humans. Perhaps as much as a third or more of our brain's information processing capability is involved in dealing with the information that arrives on waves of light—or if you wish, particles of light. As the quantum physicists have shown, it's all the same. But a third is not all. We also learn about the world around us and within us through our other senses. They include smell, taste, hearing, touch/pressure, and touch/temperature. The equally important sensory detectors within our bodies tell our brain about the positions of our bones and muscles, how our body is bent or moving, and by means of pain signals whether something is ailing or broken.

All these senses provide information to our brain. And it is with our brain that we create an internal map or image of the world outside. That internal image or map is the basis of our imaginative vision of reality.

In 1991, several years after the Loma Prieta earthquake, I had come to Santa Cruz, California, to interview two scientists doing research on the mechanism of human sight. In addition to physical sight, however, I also learned something about the many-layered nature of imaginative vision.

My wife and I had driven into town and visited the waterfront. The beach, the amusement park, and the people strolling the boardwalk all seemed to look pretty normal. But another reality lay close to the surface. Broken buildings and bulldozed city blocks revealed that the town had still not recovered from the earthquake.

As we walked about and had lunch, I found myself plunged into still another layer of inner vision, of imaginative reality. For in my mind and memory I was reexperiencing Santa Cruz as it had been in the mid-1970s. My first visit there had been with the woman who later became my first wife, Marie Celestre. On our way to San Francisco for the Christmas holidays, we stopped in Soquel, a small town near Santa Cruz, to visit Marie's childhood friend Susan. The visit afforded us an opportunity to see Santa Cruz's famous boardwalk and amusement park. I had been enchanted by the mingled smells of hot dogs and suntan oil, the sounds of people screaming on the rides, and above all by the sights—of hippies and wannabe hippies, bookstores bursting with ideas disguised as books, and boutiques crammed with colorful clothing. . . . All that was more than fifteen years in the past, but the eye of my imagination saw it clearly. That first visit could have happened just the day before.

There was nothing "scientific" about this multilayered imaginative vision of the far or near past, some personally experienced and some lived only vicariously through TV news reports. But it is from such vivid experiences, stored in memories and then forged in the fires of imagination, that both revolutionary poetry and revolutionary science are made.

When Albert Einstein was still a child, he tried to imagine what the world would look like to him if he were riding on a beam of light. No one has ever done such a thing, nor is such a feat possible. No material object can travel at the speed of light. But that's not the point. Like you or me, Einstein had experienced looking at the scenery while moving. Perhaps, for him, it was while riding in a carriage. For us, it might be riding in a car, or on a train, or looking out the window of a 747 as it rises from the runway. Einstein could draw from a wealth of sensory experiences stored as memories in his brain. And he could apply those experiences to a "what if" question: "What if I could ride on a beam of light? What would the world look like?" With that imaginative "what if" as his springboard, he would eventually leap to a new mathematical vision of reality. Today we call it the special theory of relativity. One of the triumphs of special relativity is that it provides an accurate picture of what happens to material objects when they are moving at velocities close to the speed of light. Einstein had thus succeeded in describing, in precise mathematical language, what the world would look like to him were he riding on a beam of light.

In the late 1940s another physicist was struggling with a related problem. Richard Feynman was trying to create a usable theory of quantum mechanics that took into account special relativity. Quantum mechanics is a theory of physics that deals with the movements and interactions—"mechanics," in physics parlance—of subatomic particles. Some subatomic particles are the building blocks of atoms. Others are the by-products of collisions between other subatomic particles. The best-known subatomic particles are protons and neutrons, which make up the nuclei of atoms, and electrons, which surround the nuclei of atoms like a cloud.

As a theory—that is, as a good working model of physical reality—quantum mechanics worked superbly well. So did special relativity. Some way must exist to combine the two. Feynman was working on such a theory, called quantum electrodynamics, or QED. This is a relativistic quantum theory that describes the properties of electromagnetic radiation and how it interacts with electrically charged particles like electrons. QED's basic equations deal with the emission and absorption of light by atoms and with how electrons interact with light. Charged particles like electrons interact by emitting and absorbing photons, which are the particles of light that transmit electromagnetic forces. QED is therefore often called the quantum theory of light.

Up to that point every attempt to create a valid theory of quantum electrodynamics had resulted in a mathematical disaster. It was like trying to divide 1 by 0. You get ∞ (infinity). At the end of the 1940s, physics was in turmoil. The inability to create a working theory of quantum electrodynamics would be a disaster for the field, for it would mean that one of these two eminently successful theories was in fact totally wrong.

In 1948 Feynman triumphed where others had failed, and in 1965 he shared the Nobel Prize in physics for his breakthrough.[1] Feynman's theory essentially allowed physicists to write an infinite sum of every possible way that two subatomic particles could interact. The particles interacted by exchanging what are called *virtual* subatomic particles. A virtual particle is one that exists for only an extremely short slice of time. Were it to exist any longer than a quantum blink, its existence would violate the law of conservation of energy, the law of physics that says that the total amount of energy in the universe stays constant. By exchanging virtual particles, that law remains in force and two subatomic particles like electrons can have a brief but meaningful relationship.

Subatomic particles like electrons, and entities like virtual photons, are far too small for us to ever see, no matter how powerful a microscope we might use. One truly striking aspect of Feynman's theory of quantum electrodynamics, however, was *visual*. He created a type of drawing that depicts the complex mathematical formulas that make up quantum electrodynamics. Today we call these drawings Feynman diagrams (see Figure 2).

Later in his life, Feynman would emphasize the profound strangeness of quantum mechanics. It deals with aspects of subatomic reality that are utterly beyond our commonplace, everyday experiences. Common sense does not apply in the quantum world. There is no way, Feynman would say, that you or I can

Figure 1. Richard Feynman, the developer of quantum electrodynamics and the winner of the Nobel Prize, used his inner visions of reality to change the way we perceive the world.

possibly imagine what these interactions are like. Mathematics is the only way to describe them.

But the truth is, like Einstein, Feynman *did* use his inner eye to deal with these profound mysteries of physics. In his lectures to undergraduate students at the California Institute of Technology, he would often resort to colorful analogies and "what if" scenarios to explain quantum physics. What's more, the Feynman diagram itself stands as a monument to the power of clear imaginative vision. The Feynman diagram is a visual image of something that literally can never be seen. It was born from Feynman's own

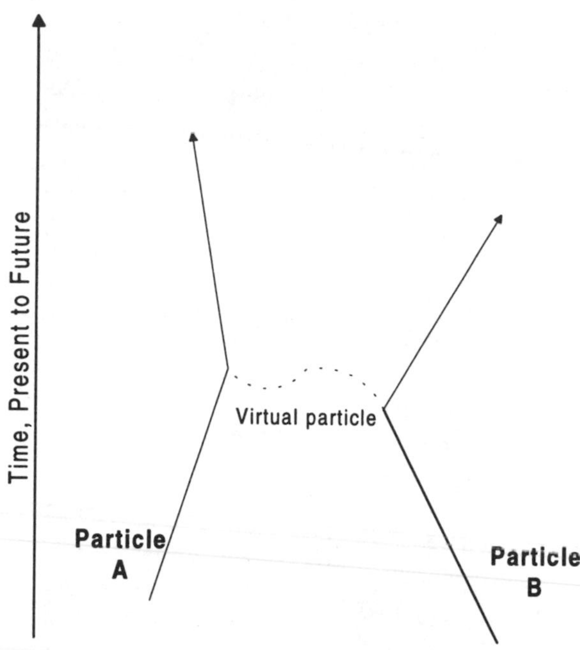

Figure 2. A simple Feynman diagram. Two electrons interact by exchanging a "virtual photon." The result is a force of repulsion, and the two electrons move apart.

imaginative playfulness about his chosen field of work, his own inner vision of quantum reality.[2]

Scientists, especially physicists and astronomers, often use the language of mathematics to express their imaginative insights. Poets use words. Had I been a poet like William Wordsworth, I could have turned my emotional and imaginative experience of Santa Cruz into a good poem. At the beginning of the nineteenth century, Wordsworth celebrated his definition of vision in a way that changed English literature. He did so partly in opposition to a way of seeing the world created by Isaac Newton more than two hundred years earlier.

In the preface to the second edition of *Lyrical Ballads*, first published in 1798 and coauthored with fellow English poet and friend Samuel Taylor Coleridge, Wordsworth proclaimed that "All good poetry is the spontaneous overflow of powerful feelings: it takes its origin from emotion recollected in tranquillity." He added, "The objects of the Poet's thoughts are every where; though the eyes and senses of man are, it is true, his favorite guides, yet he will follow wheresoever he can find an atmosphere of sensation in which to move his wings."[3]

This was a controversial position to take in the literary world of the early nineteenth century. The then-contemporary literary position placed great emphasis on form and on the role of the cool, rational intellect in poetry. Wordsworth, however, proclaimed the supremacy of the inner eye, of the poet's imaginative vision inspired by direct experience. Undaunted by the hostile response his position invoked, he continued to write poems that followed his principle, creating his poems from the recollected sights that played upon the screen of his memory.

Wordsworth, Coleridge, John Keats, Lord Byron, Percy Shelley, and several other English writers of the late eighteenth and nineteenth centuries were leaders of the literary and artistic movement known as Romanticism. It was a reaction against classicism[4] and the philosophy of rationalism, with its emphasis on reason and order.

Rationalism and classicism, in turn, had risen in large part from the remarkable scientific discoveries of Isaac Newton. In 1665 he deduced the scientific theory of universal gravitation. The story that Newton formulated his theory of gravitation after watching an apple fall from a tree was actually first spread about by the French writer Voltaire, who claimed he heard it from Newton's stepniece.

In his paradigm-creating book *Principia Mathematica* (finally published by his friend Edmond Halley in 1687), Newton laid out his discoveries about gravity and the nature of the universe. From his discoveries sprang an imaginative image of reality that held sway in Europe and later America for more than two and a half centuries. It was a vision of a universe of *order*. The great law of gravitation ruled over all: The gravitational force between any two bodies is directly proportional to the product of their masses and inversely proportional to the square of the distance between them.

Here on the surface of Earth, Newton's three laws of motion governed the behavior of moving bodies: (1) A body at rest tends to remain at rest, or a body in motion tends to remain in motion at a constant speed in a straight line, unless acted on by an outside force; (2) the acceleration of a mass by a force is directly proportional to the force and inversely proportional to the mass (increase the force and you increase the acceleration; increase the mass and you *decrease* the acceleration); and (3) for every action there is an equal and opposite reaction.

The philosophy of rationalism and the aesthetic movement called classicism are the direct offspring of this orderly vision of the cosmos inspired by Newton's discoveries. Alexander Pope was the epitome of the English classicist writer. In his poem "Epitaph Intended for Sir Isaac Newton in Westminster Abbey," he wrote:

> Nature and nature's laws lay hid in night;
> God said *Let Newton be*! and all was light.[5]

In contrast, the Romantics generally exalted nature and believed in the innate goodness of human beings, admired individuality and imagination, and celebrated the senses and emotions over reason and intellect. They were fascinated by the medieval, exotic, primitive, and nationalistic. Though English literary Romanticism formally began with Wordsworth's and Coleridge's *Lyrical Ballads*, William Blake's mystical poetry, drawings, and paintings foreshadowed the movement. All the literary and artistic Romantics of the eighteenth and nineteenth centuries celebrated the supremacy of the eye of the imagination. Raw experience was the clay with which they worked, and the furnace of their imaginations was the kiln in which they fired it.

The Wellsprings of Vision

Romanticism may not be in vogue today, but Wordsworth's philosophy of poetry had more truth to it than he ever imagined.

Just as Coleridge or Wordsworth created poetic realities out of recollected experiences, so each of us can be said to weave our visions of reality from the threads of direct experience, recalled instants, minutes, days, or years later.

From the neurobiological processes inside our brains that interpret and store the information coming through our eyes, ears, taste buds, noses, and skin, each person fashions a unique inner vision of the world. But that vision is not experienced in isolation. We humans live in families, which are part of communities, which are part of societies. We share our individual experiences by direct word of mouth within the family and among friends and acquaintances. More than that: We have invented written language and various art forms. We use these to preserve our individual and collective experiences and visions and pass them on to larger audiences. Those audiences extend through time as well as space. When I watch Shakespeare's *As You Like It*, I am experiencing the creative vision of a man who's been dead for more than 380 years. When I read the *Odyssey*, I see in my mind's eye the mythological journey of Odysseus from ancient Troy to Ithaca, an epic poem of adventure and fate first written down at least twenty-seven hundred years ago. My father's stories about the torpedoing of the U.S.S. *Minneapolis* during the battles for Guadalcanal are not only part of our family history, but also color my secondhand experience of World War II. In my mind's eye I see him scrambling up the ladder to the deck, watching enemy planes fly overhead, hearing explosions and the sound of rushing water. I can see the mattresses stuffed into the hole in the hull, just above the waterline, and the shattered remains of the ship's bow. I try to imagine what it must have been like to sail across the Pacific Ocean and back to Pearl Harbor with a bow constructed of tree trunks.

The books we read, the TV shows and movies we watch, the plays, concerts, and operas we attend—even the foods we prefer to eat—are all individual and collective experiences we have woven into our culture. These shared visions, internalized and reprocessed in our memories, ultimately create our collective vision of reality. It is not surprising that other societies do not see the world as we do. The individual and collective sensory experiences of Japanese, Egyptians, Russians, and Australian aborigines are far different from those of Europeans or Americans. Nor should we be surprised if our vision of reality at the end of the twentieth century is in many ways radically different from that of our grandparents and great-grandparents. What a difference just a hundred years has made in what we know and how we see the universe!

For Wordsworth and the Romantics, "seeing" and "vision" were more than simply the physical sense of sight. They were metaphors for a deeper experience. In his poem *The Two-Part Prelude,* Wordsworth describes the main character's depression following the death of her children and the disappearance of her husband:

> I have heard, my friend,
> That in that broken arbour she would sit
> The idle length of half a sabbath day—
> There, where you see the toadstool's lazy head—
> And when a dog passed by she still would quit
> The shade and look abroad. On this old bench
> For hours she sat, and evermore her eye
> Was busy in the distance, shaping things
> Which made her heart beat quick.[6]

Wordsworth expert Duncan Wu says of this passage that "Margaret [the character sitting in the arbor] is engaged *in an imaginative act, as her mind shapes what she sees in the far distance* in the hope that it might be her long-lost husband. . . . *Although her eye has not detached itself from the rest of her body in order to travel to the far horizon, it might have done so,* so intense is her unsatisfied craving for what she has lost (italics added)."[7] Thus Wordsworth uses the physical facts of vision, light, and seeing as metaphors for "an imaginative act," an act of "seeing" that reaches far beyond the neurological workings of the brain or the physics of light. It is an inner vision, re-created in memory from the seed of an earlier sensory experience. Margaret in *The Two-Part Prelude* is a stand-in for Wordsworth, for you, for me. And perhaps, as we'll see later, for astronomers as well, who also have eyes "busy in the distance, shaping things / Which [make their hearts] beat quick."

However, the "imaginative eye" of Wordsworth and Keats, the "inner vision" that they so dearly loved and that they felt was the wellspring of their poetry, was not merely the sense of sight. They knew quite well that the entire physical sensorium[8] provided the raw material for their craft. A celebration of and love for nature was a hallmark of the Romantic poets, and they reveled in every sensory impression. From sight to smell, from touch to taste to sound, all the senses participated in providing triggers for the "powerful feelings" later "recollected in tranquillity."

Wordsworth, in particular, had a highly tuned inner ear. In "Lines Composed a Few Miles above Tintern Abbey," for example, he invokes sound:

> again I hear
> These waters, rolling from their mountain-springs
> with a sweet inland murmur.[9]

In the Introduction to *The Prelude* he depicts the falling of an acorn

> from its cup
> Dislodged, through sere leaves rustled, or at once
> To the bare earth dropped with a startling sound.

In "Cristabel," a poem about spiritual seduction in a medieval castle that relies heavily on visual images, Coleridge also evokes taste with the lines:

> I pray you, drink this cordial wine!
> It is a wine of virtuous powers;
> My mother made it of wild flowers.[10]

And in "The Rime of the Ancient Mariner," Coleridge conjures up the smell of dead, rotting flesh by denying its presence:

For the sky and the sea, and the sea and the sky
Lay like a load on my weary eye,
And the dead were at my feet.

The cold sweat melted from their limbs,
Nor rot nor reek did they:[11]

But more than any other sense, it is vision that the Romantics invoke in their poems. In *The Prelude*, Wordsworth speaks of how

Magnificent
The morning was, in memorable pomp,
More glorious than I had ever beheld.
The Sea was laughing at a distance; all
The solid Mountains were as bright as clouds,
Grain-tinctured, drench'd in empyrean light;
And, in the mountains and the lower grounds,
Was all the sweetness of a common dawn.

In "Eve of St. Agnes" John Keats, in a few well-chosen words, paints a vivid picture of the Beadsman saying his rosary,

while his frosted breath,
Like pious incense from a censer old,
Seem'd taking flight for heaven,[12]

George Gordon, Lord Byron, in *Childe Harold, Canto Three*, conjures up an emotion-charged vision of the sea:

Once more upon the waters! yet once more!
And the waves bound beneath me as a steed
That knows his rider. Welcome to their roar!
Swift be their guidance, wheresoe'er it lead!
Though the strain'd mast should quiver as a reed,
And the rent canvas fluttering strew the gale,
Still must I on; for I am as a weed,
Flung from the rock, on Ocean's foam to sail
Where'er the surge may sweep, the tempest's breath prevail.[13]

Poets use visual images. Scientists like Feynman call upon vision—real and internal—to create new ways of understanding levels of reality we literally cannot see. All of us spin visual metaphors from our daily lives and together weave a cultural reality of symbol and transcendent meaning. The metaphors of sight and vision that we have created to describe reality are rooted in the visual information that comes to us through the biological sense organs sitting on either side of our noses. The eyes are a marvel of evolution, sophisticated organs for perceiving the tiniest amount of light and quickly sending the signals to the brain.

The preeminent role of vision is reflected in the brain, which must process and interpret the data the eyes provide. In both birds and primates, two classes of vertebrate animals that rely heavily on vision, the visual areas of the brain are much larger and more complex than the areas devoted to other sensory systems. From the large visual cortex at the back of the head to the processing centers within the cerebral cortex (the "new brain" of *Homo sapiens*, with its characteristic folds), the human brain is very much a vision-involved organ. Up to 30 percent of the human cerebral cortex is involved in processing information coming from the visual cortex (Figure 3). This is not to say that our other senses—smell, taste, hearing, and touch (which is actually a complex of several sensory inputs, including pressure, temperature, and texture)—are insignificant. They are not. Nonetheless, sight surmounts all, and by a wide margin. The auditory nerve, for example, contains about thirty thousand nerve fibers running from the ear to the brain. The optic nerve, by contrast, contains some two *million* nerve fibers—sixty times as many as the main conduits for hearing.

The human visual sensory system includes the eyes, the optic nerves and tracts, an area within the brain called the optic thalamus, and the visual cortex. But the cerebral cortex also plays a vital role in vision. This is the outermost part of the brain, with its wrinkles and folds, divided into the left and right cerebral hemispheres. Areas in the cerebral cortex take the data our visual sensory system provides and use them as raw material to reason, symbolize, daydream, and fantasize.

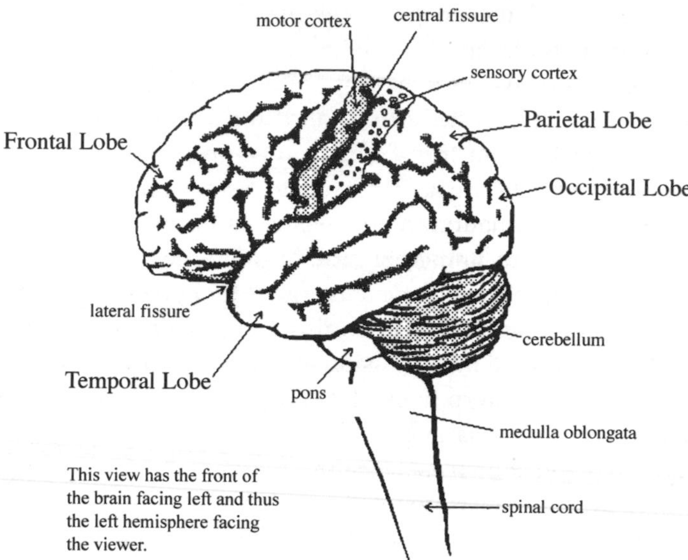

Figure 3. The main parts of the human brain are shown in this illustration. The cerebral cortex plays a vital role in the creation of our visions of reality. (Illustration from Microsoft Bookshelf 95 [Redmond: Microsoft Corporation, 1995]. The Concise Columbia Encyclopedia *is licensed from Columbia University Press. Copyright © 1995 by Columbia University Press. All rights reserved.)*

Imagine that you are a man or woman living thousands of years ago. You look out the door of your hut onto a winter scene. You see:

- Clusters of bright red berries tipping the branches of a mountain ash, topped with a dusting of snow.
- The gray and brown and white branches of the mountain ash.
- The trunk and snow-lined limbs of the maple tree, a few leaves still hanging on.
- A few extended branches of the horse chestnut tree at the front of the field beyond the hut.
- The brown and white wall of the neighbor's hut.
- Snow on the field and the other huts scattered about.

- Beyond, a tangle of bare tree branches against a bright blue late-morning sky.

All this you see. The light brings the data to your eyes. Your visual sensory system transforms it into electrical impulses that travel to your visual cortex. From there the intricate interconnections of millions of neurons take over. Years of images and memories, of experiences immediate and recollected, individual and collective, have reshaped the images into something more than just sensory images. They have also become metaphors and symbols that you use to overlay some pattern of meaning onto raw perception. Perhaps you live in a society that worships a supreme being that is female rather than male, as Mother rather than Father. What you see that morning may invoke an inner vision in which

> *The ash's berry clusters are like bright red drops of blood. The maple's last leaves look alone and forlorn, barely hanging on until the next snowstorm blows in. The branches of the trees are a tangled brown and gray net against the blue sky. They make a living spider web of tree and sky, darkness and light. The trees are branches of the World Tree, whose roots hold the entire world together and whose branches capture the Sky and hold it close. When the wind blows through the trees, carrying snow and shrouding the land in white, it is like the voice of the Great Mother who whispers to us: "All things move in spiral dance, from Spring's promise to Summer's heat, from Autumn's harvest to Winter's death, and again to Spring. Those who have died still live, and their deaths will bring new life. Spring will come again. There is beauty in this white shroud, in the bare branches, in the cold sunlight splashed across the snow. Look within. Rest."*

The raw images are just that—images. Those images connect with others you carry within your brain, ideas and symbols that you share with the others of your community. This is not, however, Carl Jung's "collective unconscious." A part of the unconscious mind shared by all humanity, the collective unconscious is supposedly the product of ancestral experience. It is Jung's explanation for the existence in many cultures across time and space of common mythic themes or archetypal ideas such as the questing hero or the dragon.[14] Rather, these metaphors and symbols are part of what Walter Anderson has called the symbolic world of social reality.[15] The archetypes of the collective unconscious will always be with us; they cut across cultures. The symbolic world of social reality, however, changes as the culture changes, dies, or metamorphoses into a new culture.

Thousands of years later, in another place, I might see a similar scene. When I lived in Spokane, I did my writing in a converted second-story bedroom. I, too, could see the brown and gray branches of a mountain ash, and its snow-dusted red berries. I saw maple and horse chestnut trees lining the street, the eastern wall of the neighbor's home, snow covering the lawns, and the bare tree branches against the sky. However, many of the symbols I carry within my mind differ considerably from those of our (not so) imaginary person of the distant past. "Snow" still says "cold" and "wet." But for me and many others in my society today, "tree branches" leads to "lace," and "red berries" says "candy" and "sweet" and "Christmas." The closest object to a "World Tree" in my spiritual mythology is the Christian cross. What I know about the ancient World Tree symbol comes from books by Starhawk and Joseph Campbell. It's true that Christmas was deliberately scheduled by the Christian Church to fall on a day close to an ancient goddess festival, the winter solstice. But today, for millions of Christians, it is still a celebration of the birth of the Son of God, and part of a somewhat different set of spiritual symbols.

A symbol is something—object, word, sound, image—that stands for something else. The "something else" is usually a complex of ideas, associations, emotions, meanings. Visual perception requires the brain to take simple visual images and transmogrify them into representations of something else. They become a rich soil from which grow metaphors and patterns.

Think of the brain as a code machine. Codes are like symbols: simple letters, words, or phrases that stand for something more complex. "10 15 5 12 4 1 22 9 19" is my name in a simple substitution code. "23 Left Cross 2" may be the second variation of an instruction for a pass play in football. The brain stores codes within its network of neurons. Visual information arrives in the brain, which matches images to codes and triggers the instructions they contain.

The result is that I "see" more than what I merely see. I piece the images into a structure with meaning that becomes transcendent. I weave them into a fabric that reveals within itself another image, a "superimage" of that which is greater than myself. I take images, visual perceptions, and create another image: an image of "reality."

The Scientists and the Poets

While most of the Romantic poets celebrated nature in all its glory, their feelings about science were somewhat more equivocal. It was at best a love–hate relationship. Wordsworth in his youth was a pantheist, a worshipper of God in nature and of nature itself. For late eighteenth- and early nineteenth-century science, he had little interest or use. Percy Shelley was alternately fascinated and repelled by science, but had a particular interest in steamboats, one of the technological innovations of his time. One telling gauge of their feelings about science (as opposed to nature) is in their poetry. Alexander Pope, the dean of the classicists, had written a glorious paean in Newton's honor. William Blake, whose artistic philosophy and poetry were precursors to the Romantics, had a somewhat different view of the man whose insights changed the way we perceive reality. In *Europe: A Prophesy*, he wrote:

A mighty Spirit leap'd from the land of Albion,
Nam'd Newton; he seiz'd the Trump, & blow'd the enormous blast!
Yellow as leaves of Autumn the myriads of Angelic hosts,
Fell thro' the wintry skies seeking their graves;
Rattling their hollow bones in howling and lamentation.

And in *Milton,* Blake sounds a warning to his readers:

Now I a fourfold vision see,
And a fourfold vision is given to me;
'Tis fourfold in my supreme delight
And threefold in soft Beulah's night
And twofold always. May God us keep
From Single vision & Newton's sleep![16]

Wordsworth also had something to say about Newton. In Book 3 of his masterwork, *The Prelude,* he wrote of the statue of Newton at Cambridge University's Trinity College:

Where the statue stood
Of Newton with his prism and silent face,
The marble index of a mind for ever
Voyaging through strange seas of thought, alone.

The lines carry not only a touch of rejection, but also a hint of fearful awe when "reflecting in tranquillity" on "a mind for ever / Voyaging through strange seas of thought, alone."

Changing the Landscape

Jonathan Swift, the early eighteenth-century Anglo-Irish writer, once called vision "the art of seeing things invisible." When he spoke of "vision" he was not talking about merely physical sight, but about the imaginative sight that Wordsworth and his colleagues would later make the centerpiece of their artistic philosophy. Swift was an artist with words, the satirist who so savagely satirized science and scientists in *Gulliver's Travels.* Ironically, his definition of artistic or imaginative vision is also a good working definition of science. Science, too, is a way of "seeing things invisible," of envisioning things invisible in as many ways as possible. It is *a* way, but certainly not the *only* way. And science itself, in its approaches to envisioning, is as many-faceted as the most brilliant diamond. It is as alive and as sloppy and chaotic and self-centered and mystical and joyous as art, or religion.

And these visions of reality change. In his book *Reality Isn't What It Used to Be,* political scientist Walter Anderson says that science "is an attempt to develop a system for the evolution of constructions of reality, and to permit a graceful exit for the dinosaurs."[17] As the discoveries and insights gleaned by scientists make their way into a culture's symbolic world, that culture's inner vision of what composes "reality" changes to make room for them. And the "dinosaurs" are those "constructions of reality" that no longer jibe with what we are now able to see, hear, and touch with our inner imaginative senses.

What is reality? Philosophy classes argue the topic, and libraries of books have been written on that subject. Still we seem no closer to any definitive answer. Perhaps there is none. We know that all humans possess the same sensory apparatus for perceiving reality (whatever it may be). My eyes, for example, see light in the same way as the eyes of a Chinese woman living in Beijing or a !Kung child in southern Africa. The visual cortices of our brains work the same way to take the electrochemical impulses coming down the optic nerve and turn them into visual perception. And the human brain—the incredible and mysterious thousand-gram ball of tissue between our ears and behind our eyes that turns sensory data into memories, dreams, and reflections—is physically the same in every human everywhere. Yet different cultures at different times and places in human history have had different visions of reality. And the same culture can and does undergo changes in its vision of reality as it moves through time. It can take a century or more, but the results are just as overwhelming as an earthquake.

A temblor like the one that shook down parts of Santa Cruz in 1989 can change a landscape in mere seconds. The survivors learn that the earth is not solid and unchanging, but can shift right beneath their feet. A similar but slower process seems to take place in our perception of reality. "Wait," some say. "The world is what it is; our perception of reality has always been this way." Not true. The changes in Western society's view of the cosmos have taken longer than an earthquake to rumble through our collective psyche. But they have shaken this century nonetheless, with some of the most powerful scientific quakes of all, truly changing the way we perceive the world around us. We need only look at the scientific "landscape" as it existed at the end of the nineteenth century. The changes since then have their origins, in a very real sense, in the imaginative visions of some highly creative scientists.

The stage was set as early as the mid-nineteenth century. Coincidentally or not, this was about the time that the movers and shakers of the artistic Romantic movement began *leaving* the stage. John Keats had died in 1821, Shelley a year later. Byron had died in the Greek civil war in 1824. Sir Walter Scott, the English Romantic novelist, died in 1832. That same year saw the passing of the German philosopher-scientist Johann Wolfgang von Goethe, whose genius embraced many fields of endeavor. Best known as the author of the dramatic poem *Faust*, he was also a renowned botanist and biologist. Goethe waged a stubborn battle against Newton's theory of light, which claimed that white light contains all the colors. Goethe argued the opposite. Physical light was also a metaphor for Goethe, who wrote in his psychological novel *Elective Affinities*, "Someday perhaps the inner light will shine forth from us, and then we shall need no other light." In fact, his dying words were, "More light!"

Coleridge died two years after Goethe. Edgar Allen Poe, the foremost representative of gothic writing in America, died a miserable death in 1849. The gothic novel and short story was an outgrowth of Romanticism, which was fascinated by the shadow side of life. As Goethe wrote in his play *Götz von Berlichingen*, "There is strong shadow where there is much light." William Wordsworth, the founder of English Romantic poetry, died in 1850. The most famous of the English Romantic painters, J. M. W. Turner, known for his striking use of light in his work, died a year later. The last of the great Romantic artists, German composer Richard Wagner, died in 1883. By then the first glimmerings of a new scientific vision of reality were already visible on the psychic horizon.

Though it may not have seemed so in astronomy. The big news in astronomy at the end of the nineteenth century had to do with a new telescope and a new moon. In 1897 George Ellery Hale had established the Yerkes Observatory at Williams Bay, Wisconsin. The observatory was home to the Yerkes telescope, with a primary lens 1 meter (40 inches) in diameter. Telescopes using lenses to focus light are called refracting telescopes, and the Yerkes telescope was the largest refractor in the world. Three years later, as the century neared its close, astronomer William Henry Pickering used another, smaller telescope to discover a new moon around Saturn. The ninth to be found orbiting that ringed planet, Pickering named it Phoebe.

This discovery and the earlier technological advance at Yerkes actually did little to change astronomy's "big picture." The vision of reality then offered by astronomy was still one of an ordered, clockwork cosmos. Moons like Phoebe and our Moon orbited their primary planets. The planets wound their stately way around the Sun. The Sun, in turn, had its own place in the Milky Way Galaxy. That location was most likely somewhere near the center of the Galaxy. The force of gravity, its nature mathematically revealed by Isaac Newton more than two hundred years earlier, ruled the astronomical cosmos. Nearly everyone agreed with the position of renowned nineteenth-century astronomer William Herschel. The Galaxy, the vast lens-shaped star cloud in which our own Sun and its planets reside, contained the sum total of all the stars, planets, nebulae, gas, and dust of the universe. The universe and the Galaxy were one and the same. It was not large, perhaps 100,000 light-years in diameter, but it was all that existed. It would be another twenty years before the imaginative visions of a handful of astronomers would begin to reshape our image of astronomical reality.

What's more, a seemingly minor scientific breakthrough in the same year that Wordsworth died would play a key role in the emergence of a new and alternative astronomical reality. In 1850 the English astronomer William Bond took the first clear photograph of an astronomical body: the moon. A minor event? Perhaps, but it was the first step toward a total revolution in astronomy. With photographic film (Bond's image, by the way, was a daguerreotype, an early type of photograph no longer used) astronomers would soon extend the physical vision of humanity to the deepest recesses of the cosmos. In fact, astrophotography would be a *direct* cause for the revolution in astronomy that has completely changed our vision of the physical universe—and our place in it.

In physics, however, the foundations of the accepted vision of reality were starting to crack. In 1881, two years before Richard Wagner's demise, two physicists tried to measure the absolute velocity of the earth with respect to the rest of the universe. Albert Michelson and Edward Morley were using an instrument called an interferometer to try and measure how the earth moves through the ***ether***. The ether was the substance that was believed to fill the universe and allow waves of light to travel. That light was made of waves had been incontrovertibly proven in 1803 by Thomas Young. But waves need a medium in which to travel, ergo, the ether. Michelson and Morley had created an elegant thought-experiment that they believed they could pull off in real life. Suppose, they said, you measure the speed of light as it travels back and forth in the same direction as the earth is moving. Then, at the same time, you measure the speed of a light beam moving perpendicular to the direction of the earth moving through space. The beam of light coming back at you in the same direction as the earth's movement will seem to be moving a little bit faster than a light beam moving perpendicular to the earth's movement. Why? Because the earth itself has moved forward a bit in the time the light beam has traveled through the ether and then come bouncing back at you from the mirror in the interferometer. In this fashion, they imagined, they could detect the earth's absolute movement in space with respect to the rest of the universe.

It didn't work. The velocity of light beams traveling both parallel to and perpendicular to the earth's motion through space were identical. It would seem that the earth was stationary in the ether, but that was simply too insane to believe. The earth certainly moved through space; either there was a problem with light, or with the ether, or both.

In 1892 Hendrik Lorentz and George Fitzgerald offered an alternative—indeed, visionary!—explanation for Michelson's and Morley's findings. Perhaps, they said (and provided mathematical equations in support) objects *contract in size* in the direction of their movement. In other words, when a dart-thrower hurls a dart at a bull's-eye, the dart's length decreases ever so slightly as it is moving toward the target. If the dart-thrower should stumble and fling it sideways, then the dart contracts along its *width* as it flies through the air, since (Lorentz and Fitzgerald claimed) it contracts in size along the direction of its movement.

This contraction would be minuscule; the unaided human eye would never detect it. But on the cosmic scale it would be enough to explain away the seemingly impossible results of the Michelson–Morley experiments. By applying the correct contraction value to the earth as it moved through the ether, you could get the speed of light for the two beams to come out the same. It was a pretty outrageous, not to say imaginative, suggestion. Within twenty years their vision of objects contracting as they move would turn out to be true. Albert Einstein would incorporate the Lorentz–Fitzgerald contraction equations into his theory of special relativity. Michelson's and Morley's failed experiments, and Lorentz and Fitzgerald's audacious explanation, would eventually create new visions of both physics and cosmology.

At the time these men were struggling with the existence or nonexistence of the ether, the Western cosmological vision hadn't changed much. One's ideas about the origins and ultimate fate of the universe depended on whether one was a Deist or a materialist. Some people liked to defer to the first cause or prime mover. The prime mover was a deity that scientists and science considered unknowable. He existed outside the universe, and after having created it had stepped out of the picture. However, many scientists saw no evidence from astronomy that the universe ever *had* a beginning. The cosmos was well ordered and essentially unchanging. Stars (somehow) passed through their evolutionary processes and eventually burned out. New stars formed (somehow) from dust and gas.

This was the essence of cosmology, the scientific exploration of the ultimate origin and fate of the universe, at the end in the nineteenth century. The worldview at that time was rightly called Newtonian, for Isaac Newton had mortared its foundations in the late seventeenth century. This Newtonian worldview had supplanted an earlier vision of reality that placed the earth at the center of everything. Now, though, the prevailing worldview was one that made humans feel like insignificant specks, meaningless cogs in the cosmic clockwork. It was a cosmology in which the universe was somehow created by the Great Watchmaker, set ticking, and then left alone. Immutable laws of nature bound together the objects within the cosmos. The universe was a quasi-mechanical entity that could in principle be totally understood—and thus controlled—by humans. As for physics . . .

At the end of the nineteenth century the Western world believed to a greater or lesser extent that there is only one reality. It is revealed to us by our senses and by the mechanical extensions of our senses that we have constructed. The probing scientific questions first asked and answered by scientists like Newton, Galileo, Kepler, and others led to a revolution in the worldview of Europe and Western society. By the beginning of the twentieth century, physics had compiled a compelling vision of reality. For the most part that image of the world—that "paradigm," as Thomas Kuhn calls it in *The Structure of Scientific Revolution*—did a very good job of explaining what our senses perceived. Mechanical laws of cause and effect governed the cosmos and all it contained—including people. In theory, and eventually (scientists assured us) in practice, even the actions and motivations of humans would be explained by rational and mechanical laws. That revolution banished all that was not perceptible to the senses of humans to the land of fairy tales and superstition. If we cannot see it, it does not exist. If some kind of image of it cannot be captured in some fashion by some instrument, it is not real. If we cannot measure it or take it apart, then it is nothing more than illusion. By the end of the nineteenth century some scientists were actually proclaiming that "the end of physics as we know it" was in sight. All the major problems of physics had been solved. Oh, there were still some niggling problems here and there, but physicists would put those to rest soon enough.

Science and the Imaginative Vision

Today a gulf appears to lie between science and poetry, science and philosophy, and science and the humanities. English scientist and novelist C. P. Snow perceptively delineated this schism in his 1959 book *The Two Cultures and the Scientific Revolution*: "Literary intellectuals at one pole—at the other scientists," he wrote, and "between the two a gulf of mutual incomprehension."[18] The distance between science and these other human endeavors is kept wide by the structure of the academic curricula in colleges and universities. But the gulf is not, in truth, as wide as we think. In a sense, it does not even exist. For poetry and physics, philosophy and cosmology all have their roots in the imaginative visions of humans.

We wonder what the world is really like. Some of us try to create meaningful images with the spoken or written word. We use powerful metaphors drawn from our own sensory experiences. Others of us turn to the language of mathematics to pencil a vision of how the universe really works. But for many who do so, the impetus also begins with an inner vision, an imaginative leap from what we know to what might be true.

Our scientific instruments bring us new information, new experiences of what some part of reality is like. Those experiences move us in different directions, give us raw material for new "what ifs." They change our inner map of reality. We conjure up a new theory that will incorporate these experiences and give shape to the inner vision. And just as the mass of a star or galaxy slightly but surely bends the light passing by it from some more distant object, so our new visions of reality subtly shift our restless search for meaning. We may look "here" rather than "there." Off we go, finding new facts and new experiences that will support or weaken our vision of reality, until someone comes along and decides to look over "there," since that area has been ignored for so long.

Something new is seen. It seems to make no sense in the context of "the standard model." Then someone looks nearby and finds another anomaly. Then another, and another. One anomaly is a fluke; two might be a problem. More than two, and "the standard model" is in trouble.

When Richard Feynman drew his diagrams and wrote down his formulas for quantum electrodynamics, the world of physics heaved a sigh of relief. Fifty years earlier, when Max Planck conjured up the idea of a "quantum," the physics community didn't quite know what to think. But Planck had also pulled physics from the brink of utter disaster. Nearly 250 years earlier, between 1664 and 1666, Isaac Newton discovered the law of universal gravity, began developing the calculus, and discovered that white light is made of every color in the spectrum. In doing so, he brought to completion a scientific revolution that had begun in 1543 with the publication of Nicholas Copernicus's book *De Revolutionibus Orbium Coelestium* and that had gathered steam in 1610 when Galileo first started using a telescope.

What is the world really like? How do we envision reality today? It is a vision far different from that of our ancient ancestors who first identified the five regularly moving stars as "planets" and thought them to be gods, or at least moving stars. It is also far different from that of Copernicus or Galileo, who lived in cultures where the Christian Church was a dominant societal and cultural force. It differs from that of Newton, who for all his scientific genius still believed in alchemy, and practiced it.

In fact, our vision of reality is not only considerably different from that of both scientists and the common man and woman of the eighteenth and nineteenth centuries. It is quite unlike the vision of reality held by scientists and laypeople of just a hundred years ago. We have discovered more facts and uncovered more various kinds of scientific experiences in the fields of physics, cosmology, and astronomy in the last fifty years than all of humanity has learned in the previous four thousand. Science, spurred on by imaginative vision, is creating for us alternative visions of reality.

2

Starry Messengers, Eyes of Glass

Out of the shadows of night
The world rolls into light;
It is daybreak everywhere.

—Henry Wadsworth Longfellow
The Bells of San Blas

If science is a way of "seeing things invisible," then surely astronomy is the oldest of the sciences. For as long as we have been human, we have looked up at the night sky and wondered what it is and what it means. The stars are unmoving in relation to one another—except for those five "wandering stars." The Sun rises and sets with profound regularity. The Moon, too, travels its predictable cycle across the heavens. Then the sudden, the unexpected, surprises us, such as the appearance of a hairy star with a long tail. It appears without warning, moves night by night across the heavens, and then disappears as mysteriously as it appeared.

How to explain these visions of the night? What are they? What meaning might they have to we who watch? What causal connections, if any, exist between the the "cosmic landscape" stretching out above our heads and the terrestrial landscape in which we live?

We have been asking these questions about astronomical reality for a very long time. As the species named *Homo sapiens*, we have existed on this planet for at least forty thousand years. For nearly all of that time we had only our unaided eyes to gather information about the heavens. An untold number of cultures and societies have been born, matured, and died during those four hundred centuries. Using only what they could learn with the naked eye, they wove a marvelously colorful tapestry of explanations for what they could see in the night sky and for the possible connections between the heavens and the earth. Less than four centuries ago we invented a tool called the telescope that vastly expanded our physical vision of the cosmos. That, in turn, profoundly changed our imaginative vision of the universe and of our place in it.

Nor has this process ended. In just the last century we have stretched our sensory capacities for perceiving the astronomical universe in astonishing new ways. Astronomers like Jerry Nelson of the University of California at Santa Cruz now use artificial eyes such as the Keck telescopes to "see" light that our physical eyes will never detect. Others can "hear" the radio signals of stars and galaxies. Astronomers even have instruments that can detect the chemicals that compose interstellar clouds; in other words, they can "smell" the cosmos. All this new knowledge has sculpted our vision of astronomical reality into shapes that our ancient forebears could simply not imagine.

Same Trees, Different Forest

Standing outside in the dark, looking up and out into forever, the view of the starry dome above calls forth a symphony of emotional responses. Do it some clear and cloudless evening, if you have the chance. Drive far out past the city lights into the countryside. Find an open space and set up your folding lounge chair or spread your blanket. And look *up*. Look up for a *long time*. Say nothing. Just look.

What do you feel?

Probably awe; peace; fear; wonder; joy. You can get overwhelmed by the size of it. You may feel like you're about to fall up and into it.

If you lie there for several hours, you will see the sky move. The stars will wheel about overhead, like cold bright dots sprinkled on the underside of an umbrella turning on its handle. The centerpoint does not lie directly overhead (unless you're lying in the snow and ice of the North or South Pole). It's off at an angle.

You're a modern, late twentieth-century person, so this view of the night sky is probably pretty new to you. So many of us live in the big cities of our society, awash in artificial lights twenty-four hours a day, that we never see the night sky as it really is. Stand in Central Park in New York City at night (if you dare), and you'll see few if any stars. The same goes for Seattle, Los Angeles, Minneapolis, Tucson, Albuquerque, Denver, or any other city of any size in America. Or, for that matter, in Canada, Mexico, South America, Europe, Australia, Africa, or Asia. Light pollution splashes across the sky and drowns the great lights beyond. You rarely see stars like this! No wonder it feels overwhelming. Perhaps you've read some books on astronomy or have gone to a planetarium show. You may recognize some of the constellations: the Little Dipper with the pole star at the end of its handle, the cup slowly turning around it; the Big Dipper; Orion, with red Betelgeuse slowly burning on his shoulder; perhaps Cassiopeia's "W." If you're viewing the sky from someplace outside Sydney, or Auckland, or Capetown, or Rio de Janeiro, you might identify the Southern Cross, or Dorado, or Sextans.

But if you were living in another culture or at another time in history, your inner vision of that cosmic landscape above your head would have been different. The imaginative landscape is composed of more than just sensory perception. It is even more than visual perception, which so dominates our sensorium. In the case of astronomy and its close cousin cosmology, though, sight has been the fuel for the fires of inner visions. For our prescientific ancestors, no sight was more immense than that of the night sky. Ten thousand years ago, the view of the heavens was incomparably better than what we have today. We humans were nomads then, living and dying in hunter-gatherer groups scattered across the face of the earth. The only lights of our own creation were fires. The night sky was an intimate companion, and we knew it well. Even a thousand years ago, we had a different view of the night sky. The giant cities of today, with their electric lights and neon signs, did not exist. Most of us lived in small towns or hamlets. We were familiar with the night and its lights. Nothing obscured it but clouds or smoke. It should come as no surprise, then, that astronomical visions figure so prominently in prescientific cosmologies.

Definitions and Requirements

Astronomy may be the oldest of the sciences—but what, exactly, is *science*? Though science certainly depends on the development of technology, it is more than simply making tools. Chimpanzees use twigs as tools, but do not do science. Paleoanthropologists, scientists who study the evolution of the human species, now have evidence of crude tools made by prehumans that goes back more than a million years.[1] But no evidence exists that these evolutionary ancestors of ours did science.

One good dictionary has defined science as: "The state of knowing," or "knowledge covering general truths or the operation of general laws esp. as obtained and tested through scientific methods," or "such knowledge concerned with the physical world and its phenomena," or "a department of systematized knowledge as an object of study" (like "the science of astronomy").[2] *Compton's Interactive Encyclopedia* defines science as "the active process by which physical, biological, and social phenomena are studied."[3]

Some of these definitions presuppose the use of measurements that produce quantifiable results. But they all assume something else, something uniquely human. No matter what our definition—"seeing things invisible" or "the state of knowing"—science is rooted in thought and therefore in language, the verbal and written expressions of our sensory perceptions and mental processes. Science is a process of explanation. It begins with our attempts to explain why the world or the universe is the way it is. To uncover the laws governing the forces that shape and run the world may offer a way of controlling those forces. If control is not possible, then perhaps prediction is.

Science is not the inevitable result of the ability to think abstractly. Neanderthal humans living in Europe and the Middle East seventy-five thousand years ago used speech to communicate with one another, created objects of art, and ritually buried their dead.[4] Cro-Magnon humans living forty thousand years ago were the first forms of modern *Homo sapiens*.[5] Abundant evidence exists of their artistic ability and immersion in spiritual rituals, and we are their direct descendents. Religious thought, spiritual constructs of reality, and creations of ritual and ceremony all require a sophisticated level of abstract thought. They require, further, the ability to communicate these abstractions through the use of spoken language. Both spirituality and its accompanying religious structures most likely arose from the desire of early humans to explain why the world is the way it is, to understand the patterns and rhythms we perceive in reality through the use of our senses, and to exercise some measure of control over the forces of nature. So did science.

This process involves a generalization from concrete sensory experience and reality to abstract concepts and symbols. No irrefutable evidence exists that any other creature on Earth today possesses such sophisticated mental abilities. Our ancestors must have begun developing them tens of thousands, and perhaps hundreds of thousands, of years ago. Certainly by the time early modern *Homo sapiens* appeared on the scene, they possessed the necessary complexity of mental processes to do science. They had been observing the natural world around them for untold hundreds of centuries. They had undoubtedly used their developing gift of language to pass on the information they learned to their children, and their children's children. From this fertile soil sprang religion and art. And, also, the beginnings of science.

Still, no evidence exists that either Neanderthal or Cro-Magnon practiced science. As an organized body of thought—the basis of science as we understand it today—science did not appear until about 600 BCE,[6] in the Greek colonies of Ionia in present-day Turkey.[7] But even before that, humans had begun developing rudimentary forms of mathematics and astronomy.

The Beginnings of Astronomical Visions

Science, at its simplest level, has to do with knowledge of the natural world. All animals have some form of instinctive ability or knowledge of regularities in the world around them. The ability to detect regular patterns and rhythms is essential to survival. Humans, too, are animals, and naturally attuned to the rhythms of nature. Add to that the extraordinary evolutionary development of our brain and a concomitant ability to think and reason in subtle abstract fashion, and you have an animal that can do a truly remarkable job of detecting regular patterns in nature.

Some thirty thousand years ago the Cro-Magnon peoples in France and central Europe were tallying numbers on stones, sticks, and pieces of bone. One wolf bone from this period has eleven sets of five cuts inscribed upon it. About twenty thousand years ago, humans living in what is today Jordan and Israel were recording sequences of numbers by making notches on bones. The nature of the numerical sequences suggests they were simple lunar calendars. The Mayans of Central America had developed astronomical constructions by about 8000 BCE. Seven thousand years ago, people in Egypt and western Asia had begun developing floodwater agriculture and could predict the arrival of the spring floods with reasonable accuracy using astronomical information.[8]

When humans began domesticating animals and plants and developing the technology of agriculture, the need for accurate understanding of natural patterns became very important indeed. Even before the development of agriculture, though, people had begun living in permanent settlements—towns and even cities. The city of Jericho, arguably the oldest continuously inhabited city on Earth, was founded before the rise of agriculture. The reason: trade. Jericho, and other towns like it, sat astride important trade routes. Various groups of hunter-gatherers and other seminomadic tribespeople had goods that could be traded. Here was a place where all could meet and swap. How do we keep track of who's gotten what in return for how many of these, anyhow? Tally them up; make marks; add and subtract. When shall we arrive again? Look to the sky; count the full moons.

The first civilizations began to appear in about 5000 BCE. In Central America, China, Egypt, India, and Mesopotamia people began living together in large cities with centralized governments and organized economies. It is then that the seeds of science were planted. This was not science as we know it today, with its philosophical underpinnings of materialism and positivism and its practical underpinnings of reproducible observation and experiment. Rather, the science of ancient times was primarily the organized study of nature's rhythms and regularities. In these earliest civilizations we begin to see a strong connection between religion and astronomy. Our ancestors took the information about the heavens that they gathered with their senses—namely, their sense of sight—and wove it into a metaphorical fabric. The visions of reality they created became, if not the foundation for, then certainly an integral part of, the structure of their religions and myths.

The Ancients' Night Sky

Astronomers today observe a wealth of phenomena in the larger cosmos: clouds of dust and gas, neutron stars, colliding galaxies, galactic superclusters that stretch like sheets of light across the universe. None of these astronomical phenomena were known even a hundred years ago, much less a thousand or ten thousand years ago. The only tools humans had to observe the cosmic landscape above them were their eyes. But even with unamplified eyesight our ancestors could perceive many fascinating objects in the night sky.

The Sun and the Moon, of course, dominated our vision of the cosmos. The Sun rose in the east and set in the west each day. As the year progressed the Sun's highest point in the noonday sky rose and fell. The days grew longer as summer approached, then receded to the same length as the night, and then in winter shrunk to a few precious hours. The Moon made a regular journey across the sky over a twenty-nine- or thirty-day period. It waxed and waned in brightness, growing from a thin crescent to a pregnant fullness, and then shrinking to invisibility. The Moon had marks or splotches on it. Some cultures saw the face of a man there; others a rabbit; still others different images.

Stars sprinkled the night sky. What they were was open to interpretation: lanterns of the gods, spirits of the dead, distant campfires of strange beings. In their nightly dance the stars wheeled about a central point that was like the tip of a pole that rose from Earth to Heaven. Some cultures identified it literally as the tip of the cosmic tent pole that held up the fabric of the night. Peoples living in the earth's Northern Hemisphere noticed that one moderately bright star lay near that central point.[9]

Different cultures saw different patterns to the arrangement of the stars in the night. Here was a camel; there a bear; over there a crown; that one a famous god or cultural hero. Their patterns would rise and set; all moved up or down in the sky as the seasons progressed. But they all stayed in the same places in relation to one another. In that sense, the stars did not move.

Well, most of them didn't move. Five particular stars seemed to wander through the heavens in somewhat more complex patterns. One was barely visible at sunrise and sunset, clinging close to the Sun. A second was a brilliant, white beacon that often preceded the Sun's arrival in the morning or followed it down in the evening. A third was a bright red wanderer that slewed through the night skies over a roughly two-year period. Two others, one yellowish and the other more white, also wound their way through the stars. One took about twelve years, the other about thirty, to make a circuit. Today we know that these "wandering stars" are the planets Mercury, Venus, Mars, Jupiter, and Saturn.

Every night people could see some stars fall from the sky. Little flashes of light, most of them were. Occasionally, though, one would blaze brilliantly, trailing a tail of sparks and fire. We now know that these are meteorites, tiny flakes of dust and occasional chunks of rock that fall from space and burn up in the earth's atmosphere.

Once every generation or so a new star would briefly appear in the sky. Usually it would blaze up at night, in the midst of some star group, last a few days or months, and then disappear. But some new stars, very rare ones that occurred only once in a few generations, would actually be visible during the day. Astronomers would later identify these as novae and supernovae, stars that either periodically flare up in brightness or—in the case of some supernovae—actually blow up.

Just as unexpected were the stars that appeared to have a tail or a head of hair. Sometimes visible at sunrise or sunset as well as among the stars, these strange heavenly creatures would slowly move from night to night against the background of the stars. Eventually they would dim from sight. But always there were others that appeared, over the years and generations of our lives and our descendents' lives. These, we know today, are comets.

Finally, arching across the night sky was a heavenly band of light. To some it appeared like a celestial river; to others a flow of milk spilled from the breasts of the goddess. It was always there, wheeling nightly around the pole, moving with the other stars as the seasons came and went. This is the Milky Way, and the Milky Way is really the disk of our home Galaxy seen edge-on from the inside.

The movements and actions of the Sun and Moon seemed to have clear connections to occurrences on Earth. These primary heavenly bodies appeared tied to the seasons, to the growth of plants and later of agricultural crops, and to the monthly flow of a woman's blood. Some peoples noticed an apparent connection between the movements of certain stars and earthly phenomena. The Egyptians, for example, detected a relationship between the annual flooding of the Nile River and the movement of the star we know as Sirius.

The wandering stars were harder to figure out. At various times a wandering star would appear to stop its constant movement in the sky, run backward for awhile, and then stop and move forward again. This is called retrograde motion, and it added to the mystery of the wandering stars. Different cultures eventually discerned what they believed to be causal connections between stars' movements and certain spiritual, mythological, or psychological events. And the "hairy stars" came to be seen as portents of evil or tragic times.

Visions of Gods and Goddesses

Like most sciences, astronomy began as a descriptive science. Early astronomers observed the night sky and described the regularities and rhythms they discerned. Even today, astronomy has largely remained descriptive. Other sciences, such as physics and biology, are today experimental as well as descriptive sciences. They use the process of reproducible experimentation as a method of uncovering truths about the universe. Astronomy involves some experiments, but not many. The subject matter of contemporary astronomy ranges from the Sun and its planets to distant stars and galaxies. It is difficult enough to carry out reproducible experiments on the planets and moons in our own solar system. It's impossible to do so with stars and galaxies. Like their ancient predecessors, today's astronomers must rely on reproducible observations as a way of uncovering astronomical truths. Reproducible observations and reproducible results are essential to real science, which seeks authentic knowledge of nature's processes. Guesses, hints, and even earnestly told stories ("I ate this herb and now I'm cured!") are not enough.

The observations of the earliest astronomers had both religious and practical aspects. For early humans, the night sky was a towering fact of life. No fluorescent street lights spattered the darkness with a white haze. The night sky was *huge* and filled with stars, a light-speckled bowl that arched over all. Contemporary science officially has nothing to do with religion or with the human impulse to spirituality. But few if any humans living thousands of years ago could deny this overwhelming reality: *Powers exist that are greater than I.* The heavens themselves were such a power. They were big, mysterious, and connected to the lives of the people.

The vastness of the heavens and the clear power and majesty of their celestial inhabitants led early humans to some perfectly reasonable (for them) conclusions and actions. They gave divine aspects to the night sky and heavenly bodies therein. As Earth was the goddess, the Great Mother, so the Sun was often the physical representation of the god, the Great Father (sometimes, though, these appellations were reversed). The Moon was a goddess, of course. Women's menstrual periods followed (more or less) the same cycle as that of the Moon. The Moon's regular waxing and waning was also a powerful image for the goddess's three great aspects of maiden, mother, and crone.[10]

Human cultures the world over came to identify the five wandering stars as representations of different gods or goddesses—or even as the deities themselves. Then there were the stars themselves. Like all animals, humans have an innate drive to identify patterns and rhythms in nature. It's a vital survival mechanism, and it is ground into the deepest structures of the brain. Even when no rhythms or patterns exist, the brain strives to find them. It will impose patterns where none exist. So it is—and has always been—with our vision of the starry night sky. Ancient humans of every culture found an image on the face of the full moon and patterns in the distribution of the stars. We call them constellations (the name comes from the Latin word *constellatus*, meaning "studded with stars"), and they are arbitrary configurations of stars in a particular area of the sky. All human cultures have created their own particular set of constellations. In ancient times, they often came to be associated with mythological heros and heroines or with different gods and goddesses.

Myths, of course, figured prominently in this spiritualization of the sky. Ancient astronomy and the mythology of a culture had intimate connections. This may seem a far cry from today's astronomy, with its orbiting telescopes, electronic CCD cameras, and immense vistas of space and time. Nevertheless, the heliocentric solar system and the Big Bang both have their roots in these ancient myths. Today's astronomical paradigm of redshifts, distant quasars, and gargantuan black holes ultimately grew out of other worldviews, visions of reality that included gods and goddesses, celestial spheres, and heavenly archers.

The heavens could be unpredictable, yes, but for the most part the denizens of the night sky followed predicable patterns. People soon realized that some of those patterns were apparently connected to earthly developments. In Egypt and Mesopotamia, the cradles of Western civilization, this connection reached into the very heart of daily life. For both civilizations the connection had to do with agriculture and irrigation.

Humans had begun domesticating plants and animals about ten thousand years ago, separately and simultaneously in Asia, the Middle East, and North and South America. In Egypt, the Nile River was the lifeblood of the community. Ancient Egypt was basically a strip of fertile land running along the banks of the Nile through the midst of the Libyan desert, plus the rich land of the Nile delta where the river emptied into the Mediterranean Sea. As summer began the Nile would start to flood, carrying north from its sources in the mountains of Ethiopia a huge load of mud and silt. As the floodwaters receded they left behind on the riverbanks their precious cargo of new fertile soil. The Sun burned in the hot, late summer sky as the people planted new crops, and clearly the Sun god watched over his subjects.

At some point people noticed an apparent connection between the renewing flood and heavenly movements. The Nile began to rise at Memphis at about the same time as the star Sothis appeared for the first time on the eastern horizon, just before sunrise. Sothis is the star we today call Sirius, and its behavior is known as heliacal rising, popping above the eastern horizon just ahead of the Sun. Heliacal setting occurs when a star falls below the western horizon just after sunset. Sirius's heliacal rising at the same time as the Nilotic floods couldn't possibly be a coincidence; the causal connection seemed obvious to people. Even today, humans look for and often "find" causal connections between phenomena that are actually utterly unrelated. It's part of that ingrained drive to detect patterns. Remember the Loma Prieta earthquake? In another time or culture the victims of such an earthquake would look for explanations in the motivations and moods of underworld gods. Today we look for strike-slip faults and moving crustal plates.

For the early inhabitants of the Nile River valley, struggling to feed the burgeoning population by means of agricultural technology, the heliacal appearance of Sothis was a sign that the Nile would soon flood and renew the earth. Sothis wasn't always accurate, though. At Memphis, the Nile waters usually began to rise around June 25 (as we would reckon the day).[11] In 3000 BCE, Sirius's heliacal rising occurred on June 22. In 2000 BCE, though, it took place on June 30. By 1000 BCE it was occurring on July 18.[12] By the time the discrepancies made the star's heliacal rising useless as a predictor, the connection had become permanently cemented into Egyptian life and religion. Sothis/Sirius continued to occupy an important place in ancient Egyptian life and traditions. Isis, the Egyptian goddess of agriculture and fertility, became associated with Sirius.

At the same time the Egyptians were developing an agricultural calendar based on Sirius, they were also devising a more straightforward time-keeping calendar based on the movements of the Sun. The Egyptians used the heliacal principle to calculate with considerable accuracy the position of the Sun in the sky. Like other peoples around the world, they came up with a yearly calendar based on the solar movements. The Egyptian calendar had 365 days. Although it started becoming inaccurate after a century or so (the year is actually 365.24 days long), it sufficed for its purposes. The agricultural calendar and the solar calendar soon fell out of synch, but the Egyptians seemed to not worry about it much.

In fact, it appears the ancient Egyptians didn't worry much about anything protoscientific, nor did they work at it. Astronomy never developed beyond the bare necessities for calendrical work. Egyptians divided the sky into ten regions, each one with a constellation. But the night sky covers a full 360 degrees from north to south and back to north, so each Egyptian "constellation" actually included 36 degrees worth of stars. To this day we have practically no idea of what their "constellations" actually looked like. They were also simply too large to be of any use for pinpointing the locations of stars. Stellar movements meant little to the ancient Egyptians; the only star whose movement had import was Sothis. The cycle of the Moon determined months, but that was about it for the Moon. As in other early civilizations, different heavenly bodies became associated with important gods and goddesses: Sirius with Isis, for example, or the Sun with the supreme god Ra. The goddess Nut was associated with the star-studded sky, and the god Qeb with the earth below. One well-known bas-relief carving from ancient Egypt dramatically depicts Nut as a naked woman arched on her fingers and toes. Shu, the god of the air, supports Nut's weight with his arms, while Qeb lies reclining below her. The Sun, Moon, and five "wandering stars" (from which, by the way, come the seven-day week), came to have religious and astrological significance. Each was thought to rule the land when in ascendance.

Ancient Egyptian astronomy was thus limited at best. The priests of the Egyptian religion used astronomical observations to calculate the rise of the Nile and to determine the beginning of the calendrical year. Protoscientific observation served only to support agricultural technology. The Egyptian climate was pleasant, and life was good. The Nile delivered life, the crops grew and were harvested, and the Sun rose and set and was not always too hot. The ancient Egyptian culture had no need to push the development of astronomy as a science. What did interest them was the afterlife. Life on the banks of the Nile was not such a bad thing, and they really didn't want it to end there. Egyptian religion eventually focused on the question of life after death. In pursuit of that preoccupation, the ancient Egyptians devised an elaborate technology of death and reincarnation. Beginning with simple attempts at embalming and below-ground tombs, Egyptian thanatology culminated in the pyramids and elaborate embalming rites. It included a comprehensive theology of the afterlife, with special rituals and ceremonies. Protoscience, including astronomy and mathematics, hit a dead end as Egyptian culture became obsessively focused on the afterlife.

Science took a different course in an area to the east of Egypt. Starting more than six thousand years ago, a series of cultures developed and died in the region around the Tigris and Euphrates Rivers near the Persian Gulf. The cultures of Babylonia, Assyria, and Mesopotamia dominated the area for centuries, and the peoples of those cultures developed a fairly sophisticated knowledge of astronomy and mathematics. The lives of the people depended on the bounty of the Tigris and Euphrates. Unlike the Nile, the two rivers of the Fertile Crescent did not easily dispense their loads of water and new topsoil. The people had to develop a complex irrigation system of dams, dikes, and canals to harness the life-giving potential of the region. Life was not handed to them on a silver platter. Storms and floods made their life considerably more insecure than that of the Egyptians. Creating and maintaining the irrigation systems required considerable technological skill. To gain and maintain those skills and that technology required a practical mastering of astronomy and mathematics. So the Mesopotamian cultures proved to be fertile ground for the growth of astronomy and mathematics.

Historians of science generally divide Babylonian astronomy into two periods. The first stretches from the earliest settlements in the Fertile Crescent to the destruction of Nineveh in 607 BCE. The second runs from the founding of the so-called Neo-Babylonian empire to the beginning of the current era. The great advances in Babylonian astronomy occurred in that first historical phase. Babylonian astronomy mixed practical observations with religious connections. Like the Egyptians and other cultures, the Babylonians had a lunar calendar of twenty-nine to thirty days. And like the Egyptians, the early Babylonians eventually had to find a way to reconcile the lunar months with the much longer solar year. They too discovered it was possible to measure the length of the tropical year with considerable precision by marking the heliacal rising of certain stars. However, the Babylonian yearly calendar had considerably more irregularities to it than did the Egyptian one. The Babylonians tried to find a heliacally rising star for each month, measuring its rising in order to make sure the calendar was properly synchronized. They didn't always succeed. The slipups were due partly to poor weather and partly to the faintness of some of the stars they used. The Babylonian lunar/solar calendar thus ended up with a lot of ad hoc adjustments.

By around 1600 BCE, the Babylonians had begun placing considerable emphasis on mathematics and astronomy. Their careful observations resulted in the first compilation of star catalogs. They carefully plotted the movements of the five wandering stars across the night skies, including their puzzling retrograde motions. The Babylonians also developed some sort of week, similar to our present one with seven days, based on the seven-day intervals of each of the four lunar phases. Each day they divided into twelve periods called *kaspus*. They also measured the distance the Sun traveled through the sky in one day. As with other societies and cultures of the times, the Babylonians also associated the various heavenly objects with gods and goddesses in their religion. Unlike the Romans and Greeks, however, the Babylonians actually identified the five planets with specific gods. It wasn't just a case of symbolism; they truly thought that each planet *was* that particular deity. For all that, the Babylonians eventually developed a rather sophisticated knowledge of astronomy and mathematics. Their knowledge and the spiritual associations of their math and astronomy would eventually have a powerful impact on an important Greek philosopher. That man, in turn, would have an influence on Western astronomy, cosmology, mathematics—and visions of reality—that extends even to our time.

The Megalithic Vision

The members of these early Middle Eastern civilizations were certainly not the only people using astronomy and mathematics. For example, many of us are now familiar with the megalithic astronomy of the Celtic tribespeople in England, Scotland, Ireland, and France, thanks to detective work of people like Gerald Hawkins and Alexander Thom.[13] The word *megalith* is based on Greek roots and simply means "very large stone." Created for profound mythological and spiritual reasons, megalithic constructions were not built to "do" astronomy as we understand it today. The astronomical observations were merely a tool in the service of a grander vision.

Stonehenge, rising dramatically from Salisbury plain in England, is perhaps the best-known example of a megalithic or stone construction. Hawkins and others have suggested the existence of several astronomically significant stone alignments at Stonehenge. Probably the best-known one runs from the center of Stonehenge along its axis to a more distant solitary standing megalith called the Heel Stone. This alignment aims almost directly at the point of midsummer sunrise. Hawkins didn't discover this alignment even though as far back as 1740 a British historian named William Stukeley had noted it. The popular notion of Stonehenge and Druidic religion springs from his eighteenth-century illustrated pamphlets. It was Stukeley who popularized the idea that ancient Druid priests had built Stonehenge and used it for all kinds of dark and nasty rituals. The only problem with this romantic-sounding scenario is that it's not true. Archaeological evidence reveals that Stonehenge was finished at least two thousand years before the classical Druidic period of the Celtic people in Britain.

Another problem with Stonehenge and astronomy is of more recent vintage. After Hawkins published his first and most famous book on Stonehenge, numerous astronomers and historians attacked the validity of many of his conclusions. Several of the astronomical alignments Hawkins claimed turn out to be far from precise. Even the best-known one, the one that involves the Heel Stone and midsummer sunrise, is far off the mark. Hawkins used aerial photographs to refine his measurements, but they still were off by at least a degree or more. These inconsistencies cast considerable doubt on Stonehenge's putative astronomical function.

However, it now appears that we know less about the archaeological aspects of Stonehenge than we first assumed. In particular, much of the megalithic monument remains unexcavated. One dramatic illustration of this serious gap in our understanding came to light in 1979. The British Post Office, which runs the phone system as well as delivers the mail, planned to lay a new phone cable along the road that currently runs near the Heel Stone. When archaeologists carried out a quick examination of the area to be dug up, they discovered clear evidence of the former presence of a second "Heel Stone." It now appears that two such stones existed sitting side by side with a small gap of about 3 meters between them. And it looks as if midsummer sunrise would have occurred on the horizon at a point right between the two stones, as seen from the center of Stonehenge. So this most famous of all megalithic monuments may indeed have functioned as a "stone observatory," providing the Celtic shamans with vital astronomical information for their rituals and their community's culture.

Just as intriguing as those of Stonehenge are the astronomical alignments of other megalithic structures. Megalithic monuments found in Scotland include those at Staca, Kintraw, Clava, Old Keig, and Fortinall. Several other important sites in England include those at Avebury, Marden, and Black Marsh. Another megalithic construction may be found in Ireland at Newgrange, north of Dublin. This is not a freestanding set of stones like Stonehenge, but rather a huge burial mound made of giant stones and fill dirt. On the morning of the winter solstice, a ray of sunlight passes through the roof box of the Newgrange megalithic tomb, down a long narrow passageway lined with stones, and onto the back of the tomb chamber. This is not strictly an astronomical site, however. The builders' intention was spiritual rather than "scientific." No one living could see this astonishing annual event. Large stones blocked any chance of a human crawling into the tomb. Only a tiny triangular slit, deliberately constructed, allowed the winter solstice morning sunlight to enter. Clearly the light was meant to be "seen" by the bones and ashes of the dead entombed in the structure.

Though not meant to be used for scientific astronomical observations, Newgrange and other megalithic structures are still important for both historical and scientific reasons. Observational astronomy in the sense we know it today, and in the sense that the earliest agriculture-based civilizations used it, may well have its origins in these structures. But as noted earlier, these megalithic monuments were not meant primarily for astronomy. Their meaning is rooted in a vision of reality we can barely imagine today. This was the worldview of humans who perceived the presence of divinity in everything they saw. They lived in a world where solstice and equinox and cross-quarter days really meant something, a world that danced to the slow rhythms of the seasons and the monthly roundelay of the Moon. Light was not wave or particle, but *light*: alive, warm, lifegiving, and kin to fire. Darkness and night danced in partnership with day and Sun. Water was alive, and mirrored the gift of life that comes from the goddess. Cold earth, too, was Her gift, and a sign of Her relentless but necessary pursuit of all living creatures. All must die and return to the ground, so that life may come again. Life and death waltzed eternally, as did water and earth, day and night, Sun and Moon, spring and fall, summer and winter.

No modern human has caught the spirit of this ancient vision of reality quite as brilliantly as a nineteenth-century Roman Catholic priest and poet, Gerard Manley Hopkins, who wrote of grandeur: "The world is charged with the grandeur of God"; of the night sky: "Look at the stars! . . . / O look at all the fire-folk sitting in the air!"; of a falcon:

> I caught this morning morning's minion, king-
> dom of daylight's dauphin, dapple-dawn-drawn
> Falcon in his riding
> Of the rolling level underneath him steady
> air, and striding
> High there, how he rung upon the rein of a wimpling
> wing
> In his ecstasy!

—and of all nature:

> All things counter, original, spare, strange;
> Whatever is fickle, freckled (who knows how?)
> With swift, slow; sweet, sour; adazzle, dim;
> He fathers-forth whose beauty is past change:

Like his near-contemporaries Wordsworth and Coleridge, and the other Romantic poets, Hopkins used words and poetic techniques of metaphor and allusion to transform sensory perception into something more. In his poems he gives voice to an inner vision of reality. It is a glimpse of "things invisible" that is rivaled today only by the astonishing visual images of distant galaxies and exploding stars captured by scientific instruments like the Keck telescopes and the Hubble Space Telescope.

The Greek Visions

Astronomy as a physical *science*, as an attempt to explain the movements of heavenly bodies in physical terms, began with the ancient Greeks. To be precise, it began with a group of Greek philosophers who lived in the city of Miletus. Located on the western coast of present-day Turkey in the ancient province of Ionia, Miletus was a Greek colony town. It was also a center of learning. The greatest of this group of philosophers was the seventh-century BCE philosopher Thales. Thales believed that the world arose not from some mythical creation event but from a physical substance. He thought that substance was water. This made sense from Thales's point of view. He and other Greeks could see with their own eyes that the world was surrounded by water—the great ocean that we today call the Mediterranean Sea. The band of stars he could see in the night sky, which we call the Milky Way, was a heavenly analogue of rivers and streams. Moreover, water was the source of life; a man can go many days without eating, but deprived of water for only a short time he will quickly die. Ergo, it was logical to conclude that the world had been created from water. Thales's philosophical descendants included Anaximander and Anaximenes. These and other philosophers of the time led the way in proposing physically based cosmologies and explanations for peoples' astronomical observations.

At about the same time as Thales was living and working in Miletus, another important Greek philosopher was born on

Samos, an island off the Turkish coast. Pythagoras is surely one of the most controversial of the ancient Greek thinkers. We know little about his life, and none of his actual writings survive; we only have what others later wrote about his philosophy and explanations for natural phenomena. Some writers suggest that Pythagoras spent a part of his early adult life in Mesopotamia, and there learned much of the practical astronomical data accumulated by the Babylonians. Surely he knew a lot of astronomy, and there is no reason to dismiss this story out of hand.

Pythagoras was a mystic as well as a mathematician. He believed that the underlying guiding principle of the cosmos was the number. For example, he discovered mathematical principles underlying musical intervals; pleasing-sounding harmonies are based on simple ratios of numbers, such as 2:1 (an octave) and 3:2 (a fifth). These kinds of mathematical relationships led him to mystical conclusions about numbers, and from these abstractions he jumped to a mystical vision of reality in which numbers were real entities that constituted guiding cosmic principles. Pythagoras combined his mystical numerology with some of Thales's ideas to create his own vision of the cosmos. At the center (he said) was Earth, a perfect sphere. Surrounding Earth were a series of crystalline spheres, each the abode of a different planet and the Sun and Moon. As each heavenly object moved along its perfectly circular path it emitted a pure musical tone. This, then, was the "music of the spheres." The stars were fixed in another, distant crystalline sphere that also moved.

Other philosophers who followed Pythagoras modified his picture of the cosmos, adding their own notions and visions. Heraclitus, for example, who lived in the sixth century BCE, thought the earth was shaped like a bowl. Leucippus, who lived in the fifth century BCE, thought it was shaped like a drum. Like Thales before them, these men were not offering mythological constructs or explanations with symbolic import. They were offering specific *physical* explanations for what others could *see* in the sky. This is what separates a scientific theory from a myth. Myth is symbolic, psychological, even spiritual. Science, based ultimately on the philosophical musings of Greeks who lived more than twenty-five hundred years ago, chooses to eschew the symbolic and spiritual.

The next important philosophical step leading to the revolution in our perception of the cosmic landscape began in the fifth century BCE with Plato, one of the most famous of all Greek thinkers. Plato's teacher had been Socrates; his most famous student was Aristotle. Plato himself suggested that the earth, the Sun, the Moon, the planets, and even the sky itself were all spheres, the perfect solid shape. The heavenly spheres moved about the earth in circular orbits. Plato's contemporary, the philosopher Eudoxus, elaborated on this proposal. He developed an elaborate system of spheres tilted at various angles to one another to explain the movements of the Moon, the Sun, and the planets. A total of twenty-seven different spheres were required in his astronomical system to explain the motions of heavenly bodies. Eudoxus probably considered his proposal to be a mathematical exercise only, and not to have any physical reality to it. Nevertheless, it stands as one of the earliest serious attempts to create a scientific astronomical theory.

Plato's pupil Aristotle knew of Eudoxus's system and adapted it with his own modifications. Aristotle, however, believed in his system's physical reality. He was truly offering a serious physical, protoscientific explanation for the movements of stars and planets as seen by observers on the earth. Aristotle's scheme had a total of fifty-six crystalline spheres and counterspheres, moving in complex patterns to explain the real-life movements of heavenly bodies.

Like most other people of his time, Aristotle knew that the earth was round and not flat. He lived in a sea-faring nation. He could see with his own eyes what happened when a ship sailed off toward the horizon. It became smaller, and as it did so the ship's main body would disappear. Then its mast would begin to fall below the horizon, until the entire ship was lost to sight. Thus it was obvious that the earth's surface was curved. Besides, the shadow of the earth on the Moon during a lunar eclipse was circular. Just as obvious from the evidence of one's senses was the fact that the earth did not move. It was stationary. If it did move, he reasoned, we'd all be flung off its surface. We are not; ergo, the earth is stationary. Aristotle also pointed to the phenomenon called parallax to prove that the earth lay at the center of the cosmos. *Parallax* is the apparent difference in motion of a celestial object when observed from two distant points. Suppose the earth moved around the Sun, for example. Then the stars would surely appear to move in relationship to one another at different times of the year. They don't; hence, we don't move. Aristotle's understanding of parallax was correct. He did make one small incorrect assumption, though. He assumed that the sphere of stars was relatively close to Earth. Stellar parallax does take place; it's just that we can't see it with unaided eyes. And unaided eyes were all that Aristotle had, and all his descendents would have for nearly two thousand years.

The Western world lost access to Aristotle's writings for more than a thousand years, but the Arabic world did not. Copies of Aristotle's work lay preserved in the great library at Alexandria, Egypt, and elsewhere. One person who knew of this work was Hipparchus of Nicaea, who lived in the second century BCE.

Like Pythagoras before him, Hipparchus had first turned to the work of the Babylonians, using their astronomical observations to construct a physical cosmology. To these observations Hipparchus added Aristotle's explanations. However, Hipparchus quickly discovered the inconsistencies that Aristotle and his associates had ignored. It was clear to anyone who carefully watched the planets, for example, that they did not in real life move on circular paths or move at uniform speeds. Retrograde motion was also not really explained by Aristotle. So Hipparchus invented a set of geometric devices to do so. His system envisioned a series of complex geometric movements called epicycles, deferents, and eccentrics. A *deferent* is a large, imaginary circle surrounding the earth. An *epicycle* is a smaller circle along which a planet travels. The center of the epicycle, in turn, moves along the deferent, circling the earth. An *eccentric* was another circle, whose center lay off to one side of the earth's center. Some heavenly bodies traveled along an eccentric.

When Hipparchus invented the eccentric, he displaced the earth from the center of the cosmos, violating Aristotle's firm geocentric principle. But he did it to make his system conform with the reality of the senses. Some planets, like Mars, move more quickly along some parts of their paths through the night sky than through others. If Mars's deferent was centered on an eccentric rather than the earth, this movement made sense.

The Ptolemaic Vision

Even after the tragic destruction of the Alexandria library, Aristotle's work survived in the Middle East. Early Islamic scientists translated Aristotle's Greek into Arabic, and from there his work eventually made its way back to Europe in the early years of the second millennium. Long before then, though, an astronomer and mathematician living in Alexandria in the third century CE found the work of Aristotle and Hipparchus and made it his own.

We know little about the life or person of the man known to history as Ptolemy. His real name was probably Claudius Ptolemaeus; he lived in Alexandria, Egypt, during the second century CE, certainly around the year 125 CE; and he wrote about astronomy. His great achievement was a thirteen-volume treatise called *Almagest*. It included a detailed exposition of a geocentric cosmology—a view of the cosmos that placed the earth at the center of everything. This Ptolemaic worldview became the astronomical paradigm for the Western world for the next fourteen hundred years.

Like every astronomer and philosopher before him, Ptolemy had only the evidence of his eyes with which to work. Visual sensory perception provided all the information available about the nature of astronomical reality. Different societies and cultures had created a variety of explanations for the evidence they could see. Aristotle's vision of cosmic reality, as elaborated on by Hipparchus, made sense to Ptolemy. It seemed to do a good job of explaining the "why" of the "what." Earth was an unmoving sphere that occupied the center of the universe. Everything else revolved around Earth: Sun, Moon, planets, and stars all rode embedded in perfect crystalline spheres.

Ptolemy's astronomical observations revealed (he concluded) that these movements were rather more complex than Hipparchus or Aristotle had realized. He therefore modified their vision of astronomical reality with an invention of his own: the equant.

To understand what an equant is, recall that a deferent is an imaginary circle with the earth at its center. The sun, Moon, and the planets each travel along other circles called epicycles. The center of each epicycle in turn travels around the earth as it moves along a deferent. Now, imagine that the earth is *not* at the center of each deferent, but is instead slightly offset from it. Next, draw a line from the earth to the center of the deferent, and then extend it beyond. At an equal distance from the deferent's center, but exactly opposite the earth, that line will intersect the imaginary geometrical point that Ptolemy called an equant. This system neatly explained retrograde motion. Ptolemy used epicycles, deferents, eccentrics, and equants to explain these and other motions of heavenly bodies.

Ptolemy's geocentric model of the cosmos conformed quite well to the observations of the planets' movements in the sky. In creating and publicizing his system, Ptolemy may have violated some of Aristotle's philosophical tenets, but he did so for a good reason. Aristotle's assertions may have sounded lofty, but they didn't match the astronomical reality that people could perceive with their own eyes. In modifying Aristotle's system, Ptolemy devised a theory—an explanation for observations of nature—that did what the best scientific theories do. It worked. By focusing on physical observations of the visible world, Ptolemy's theory offered a deeper imaginative vision. It was an alternative astronomical reality, based on but different from that of Pythagoras, Heraclitus, Eudoxus, and Hipparchus. It was an alternate vision of reality that held sway in the West for over fourteen hundred years.

Slippage and Rebound

In 476 CE the Roman Empire in the West finally collapsed for good. (The Eastern Roman Empire, usually called the Byzantine

Empire, continued until the fifteenth century.) The next thousand years have been popularly dubbed the Dark Ages. Though not entirely accurate, the appellation does have some basis in fact. With the Empire's collapse the economic and political map of Europe went through some wrenching changes. But the time was far from a precipitous slide into intellectual ignorance and savagery, for one institution remained strong, and even gained political power: the Christian Church. And the Christian worldview, based on the Bible and the writings of various church fathers, held sway throughout much of Europe. The Christian monasteries, especially those of the Benedictine order, were repositories of knowledge during the four centuries or so following the fall of the Roman Empire. However, much of the West's intellectual heritage had been lost. In particular, no one in Europe had access to the writings of Aristotle.

Around the beginning of the second Christian millennium, whatever intellectual "darkness" that had previously held sway began to dissipate. In 1126 Adelard of Bath translated Al-Khwarizmi's *Astronomical Tables* into Latin, along with another book by Al-Khwarizmi about arithmetic. About fifty years later the Italian scholar Gerard of Cremona translated Ptolemy's *Almagest*, the great multivolume work of astronomy. In 1202 the Italian mathematician Leonardo Fibonacci introduced Arabic numerals to Europe, and they eventually replaced Roman numerals. The first of Europe's great universities began to appear: Siena in 1203, Vicenza (1204), Salamanca (1214), and Naples (1224). In 1217 Michael Scot introduced Europe to Aristotle's model of astronomy when he translated the *Liber Astronomiae* from Arabic into Latin. The thirteenth century was the time of Friar Roger Bacon, the most influential protoscientist and natural philosopher of his age.

The astronomy of Western society during the Renaissance, roughly from 1450 to 1650, was far from primitive. In fact, at least one important philosopher of those times anticipated the paradigm shift later wrought by Nicolaus Copernicus, Johannes Kepler, Galileo Galilei, and Isaac Newton. Around 1440 Nicholas Krebs, better known as Nicholas of Cusa, published a book entitled *De Docta Ignorantia* (*On Learned Ignorance*). In it he proposed the ideas that the universe was actually infinite in extent, that all heavenly bodies were pretty much like the earth or the Sun, and that the earth revolved around the Sun and not the other way around.

Nicholas of Cusa's position was very much a minority one. Most philosophers, theologians, and natural scientists accepted the prevailing cosmic worldview, that of Ptolemy's *Almagest*, which had codified the astronomical and cosmological systems of Aristotle and Plato. Despite the essential falsity of Ptolemy's geocentric worldview, astronomers and astronomy continued to advance quite handily. In 1471, seven years after Nicholas of Cusa's death, the German astronomer Regiomontanus built an observatory in Nuremburg. Thc following year he carried out a careful scientific study of a comet—the first person to do so. Today we know that the comet he studied was Halley's Comet. That same year, 1472, Georg von Peuerbach wrote *Theoricae Novae Planetarus* (*New Theory of the Planets*), basically a rewriting of the *Almagest*. In 1472 Regiomontanus's *Astronomical Ephemerides* appeared, which listed the positions in the sky of various heavenly bodies. Thirty-two years later Columbus would use Regiomontanus's ephemeris to frighten a group of Native Americans by correctly predicting a lunar eclipse.

Slowly and quietly the Western world had begun accumulating the knowledge and the tools necessary for the creation of astronomy as we know it today. Those tools and that knowledge would inevitably influence what a new generation of astronomers would observe and how they would carry out those observations. The new information they would acquire about the heavens would in turn profoundly change the inner visions of reality of a few men. And the imaginative visions of those men would eventually create new alternative visions of reality.

The single most important tool for this sea change was the telescope.

Jannsen's Tube

A telescope is, simply put, an instrument for viewing distant objects. It collects radiation of some sort from the object and brings it to the eyes of the observer in such a way as to make the object appear larger. The distant object may be producing the light, like a star. Or it may, like a planet, be reflecting light from some other source. Traditionally, scientists and others have used telescopes to directly view a distant object or to make some kind of image of it. Many astronomers today use specialized instruments, attached to a telescope, that do not actually produce a "visual image" of the planet, star, or galaxy. These instruments instead extract specific kinds of information carried in the light like a letter is carried in an envelope. What the scientist "sees" is not a magnified image of Mars, or of the star Sirius, or of the Andromeda Galaxy. Instead, the scientist often ends up with some kind of graph, chart, or matrix of numbers that depicts the data in a very nonvisual fashion.

We humans perceive the world around us predominantly through our sense of sight. Even an experienced astronomer can find "reading" data in these forms to be tedious. The recent development of powerful computers and graphics software has made it possible to turn even these collections of numbers into new kinds of images that are easier to understand.

Optical telescopes gather radiation from the visible part of the spectrum. Other kinds of telescopes detect electromagnetic radiation in other spectral regions. They include radio telescopes and infrared telescopes, which gather radiation with longer wavelengths and lower energies than visible light; and ultraviolet, X-ray, and gamma-ray telescopes, which detect radiation with higher energies and shorter wavelengths than visible light (see Table 2–1).

Still other telescopes being used or planned today will not sample the electromagnetic spectrum at all. Instead, they will collect other types of subatomic particles or waves such as neutrinos or gravitational waves.

Infrared radiation is heat radiation, and we detect heat not with our eyes but with our skin, as part of the sense we call "touch." So in a sense, infrared-detecting telescopes like the giant Keck telescopes in Hawaii are extensions of our sense of touch. We are "feeling" the heat of distant stars. In a somewhat analogous sense, radio telescopes are giant "ears." We use them to "hear" the

Table 2–1. Some Examples of Electromagnetic Radiation and Their Typical Wavelengths

Example	Wavelength (meters)
Cosmic rays	$< 3 \times 10^{-15}$ (3 femtometers)
Gamma rays	3×10^{-14} to 3×10^{-13}
X rays	3×10^{-13} to 3×10^{-10}
Ultraviolet light	3×10^{-10} to 3×10^{-7}
Visible light	3×10^{-7} to 3×10^{-6}
Infrared light	3×10^{-6} to 3×10^{-4} (0.0003)
Microwaves	0.003 to 0.03
Radar	0.03 to 0.3
TV and FM radio waves	3
Shortwave radio waves	30
AM radio waves	300
Long-wave radio waves	3,000

shouts and whispers of objects as near as the planet Jupiter and as distant as quasars lying at the edge of the observable universe.

These particular kinds of telescopes are relatively new to astronomy—the first radio telescope was built in the 1930s. The first telescopes, and the ones with which most of us are familiar, were optical telescopes.

The first description of an optical telescope appeared in the Netherlands in October 1608, and within a year the instrument was being used for astronomical purposes. However, the materials and technology to make a telescope had all been around for a long time before that. Key to the development of the telescope was the glass lens. A lens is a piece of glass whose shape deflects light from a straight path through it. This property is called refraction. Refraction occurs whenever a beam of light passes through the boundary between two media, such as air and glass or a vacuum and air. The exact amount of bending depends on the second medium's refractive index, also called its index of refraction.[14]

A convex (or convergent) lens is always thicker at its center than at its edges. It refracts incoming parallel light rays to a focal point in front of the lens. The distance from the midpoint of the lens to the focal point is called the lens's focal length. A concave (or divergent) lens, by contrast, is always thinner at the center than at its edges. When parallel rays of light pass through a concave lens, they diverge or spread apart. This makes the light rays appear as if they are coming from a distant focal point behind the lens (Figure 4).

Glass itself had been made in Egypt around 3500 BCE. Archaeologists have found crude glass lenses in Crete and Turkey that date from around 2000 BCE. In the first century CE, the Roman writer Seneca noted that a spherical bowl filled with water—a crude convex lens—acted as a magnifying glass. The Arab scholar Al-Hazen wrote about his experiments with glass lenses in the eleventh century. Roger Bacon, the English natural scientist and monk, discussed lenses in his book *Opus Majus*, written in 1267 and 1268. In a scientifically prophetic moment he wrote, "From an incredible distance we may read the smallest letters . . . the Sun,

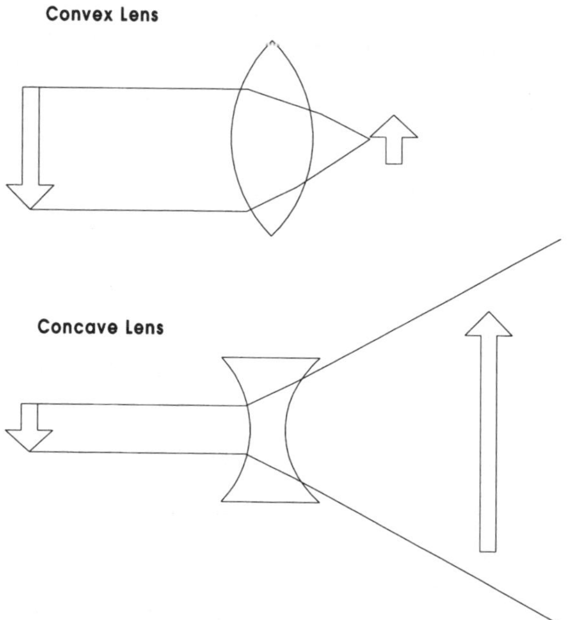

Figure 4. The difference between convex and concave lenses.

Moon, and stars may be made to descend hither in appearance . . . which persons unacquainted with such things would refuse to believe." Bacon's book was not published until 1733, but it may well have been available to a few people in some sort of medieval samizdat form. By the seventeenth century convex lenses as we know them today, deliberately ground from a solid piece of glass, had been available in Europe for at least three hundred years, and concave lenses for one hundred and fifty years. Anyone during that time could have taken two convex lenses or a convex and concave lens, put them at different ends of a stiff tube, and invented the telescope. Before 1608, though, no reliable records exist of anyone doing so. Perhaps someone did—perhaps many people did—but the invention did not catch on.

We really don't know why, but we can venture some informed guesses. For one thing, most lenses made before the seventeenth century were of poor optical quality. Glass-making was not the most sophisticated of arts at that time, and it was probably very difficult to make glass that was optically clear enough to be useful as magnifying lenses. What's more, there may not have been any pressing social or economic need for telescopes or magnifying glasses.

In the fifteenth and sixteenth centuries two important events occurred that would change all that. Around 1455 Johannes Gutenberg launched the age of the printed word when he produced the Mazarin Bible. By the beginning of the seventeenth century printed books were becoming commonplace. Many were printed in rather small type, which created a need for spectacle lenses that would magnify the size of the type for scholars and others trying to read the printed material. Eyeglasses were already fairly common in Europe by the beginning of the fourteenth century, even though their poor optical quality would have made most of them more of a hindrance to reading than a help. (Roger Bacon had also predicted the invention of spectacles in his *Opus Majus*.) The proliferation of books and other printed material simply further increased their need.

One of the books making the rounds in Europe in the mid-sixteenth century had been written by a Polish canon (in the

Middle Ages, a priest or member of a religious community) named Nicolaus Copernicus. It introduced an idea that, combined with the telescope, would eventually reshape Western society's vision of reality. Copernicus had died in 1543, and his new vision of the cosmos had been published when he was on his deathbed. In *De Revolutionibus Orbium Coelestium* (*On the Revolution of Heavenly Bodies*) Copernicus asserted that the universe was a heliocentric system. The earth did not lie at the center of the cosmos, he wrote; rather, the earth and the other planets circled about the Sun. A Lutheran minister named Andreas Osiander had overseen the actual publication of *De Revolutionibus*, and he recognized the potentially blasphemous nature of what Copernicus was suggesting. Osiander therefore added a preface that essentially said, "Actually, this isn't *real*—but it makes the mathematical calculations easier if you *pretend* it is."

The other important development occurred in 1572. That year Al-Hazen's works were translated into Latin, thirty-six years before the appearance of the first telescope in Europe. It is possible that the availability of Al-Hazen's writings on his experiments with lenses, coupled with the ever-increasing need for lenses of reasonable optical quality, set the stage for the "invention" of the telescope.

The telescope was not "invented" in the sense that Alexander Graham Bell invented the telephone or William Shockley and his colleagues invented the transistor. We know of the causal connection between these particular individuals and those particular machines. Not so with the telescope. It was probably invented, forgotten about, and reinvented many times over the centuries. Most history books assert that a spectacle-maker named Hans Lippershey invented the telescope in 1608 in the Dutch town of Middelburg. However, all we really know is that Lippershey was *making* telescopes in 1608. He sold several to the United Netherlands government that year and subsequently requested a thirty-year "privilege," or patent, on the device. The government denied his request on the grounds that the telescope was already a well-known device that several other people were building.

One such telescope-maker, often cited as the "real" inventor of the telescope, was a shady character named Zacharias Janssen. Some historical sources suggest that Janssen built a telescope in 1604, having copied the design from one that came from Italy. During the next several years he sold his telescopes at fairs in Germany and the Netherlands. Before he went into the telescope-making business in Middelburg, Janssen had been in the money-making business. Literally. He was a forger who made a tidy sum printing and distributing ersatz Spanish bills. Spain and the Netherlands were at war during this time, so Janssen's activities were not only tolerated but considered patriotic. When the war ended, however, Janssen apparently forgot to stop printing Spanish money. At that point his patriotic activities became nothing more than illegal forgery. Janssen was forced to flee Middelburg, and from there he disappeared into the mists of history.

No matter who "invented" the telescope—Lippershey, Janssen, or some anonymous spectacle-maker in some Middelburg alley—by early 1609 the device had spread beyond the United Netherlands. Any maker of eyeglasses could easily grind the right kinds of lenses and fit them into opposite ends of a paper or lead tube. And anyone with a bent for making a profit could sell the things. Within a matter of months telescopes had become the toy of choice. They were on sale everywhere in Paris. In England, Christopher Tooke constructed dozens of them. They spread to cities and towns throughout Europe—including the Italian city of Padua. There the new technology of the telescope fell into the eager hands of a hot-tempered Italian who had once considered becoming a priest: Galileo Galilei. When that happened, the telescope became more than a toy. It became an instrument for revolution.

The Man from Pisa

Galileo Galilei was born in Pisa, Italy, on February 15, 1564, the son of a well-known musician who was originally from Florence. Galileo's family was fairly well-off financially, and when he was ten his family returned to Florence. A year later Galileo began studying at a monastery near the town of Vallombrosa. At one point he even became a novice in the religious order that ran the monastery, but eventually decided that the priesthood was not his vocation. Galileo was more interested in medicine.

Nothing about the young Italian at this point suggested that he was anyone special. He was simply one of many young Italian men of the late sixteenth century, born into a middle-class family in an era when the middle class was gaining some economic clout. He could have become a priest—he had the connections. He could have gone on to be a *dottore*—he had the brains. Instead, Galileo would become the grandfather of modern astronomy and physics; he would take a small optical instrument that was mostly a toy and turn it into a machine that would overturn the way humans perceive reality.

When he was seventeen Galileo left the monastery of Santa Maria to study medicine at the University of Pisa. After two years, however, his interest in this profession had also waned. He had seized upon a new fixation: mathematics and geometry. He left Pisa and returned home to Florence, spending several years teaching and writing. In 1589 he accepted a position at the University of Pisa as a professor of mathematics. It was back to Pisa for three years.

In 1592, his appointment at the University over, Galileo took a job in Padua teaching astronomy and geometry. By then he had heard about Copernicus's new vision of the nature of the universe. *De Revolutionibus* had not exactly been a best-seller. A few hundred copies were printed just before Copernicus died, but they carried a high price, sold slowly, and soon went out of print. A second edition appeared in 1566, and a third in 1617. However, many European intellectuals were rather fascinated by Copernicus's theory, and the few copies of the book circulated among them during the years following its publication. At some point before the beginning of the seventeenth century, Galileo read a copy of *De Revolutionibus*.

Fascinated, he plunged into a study of this new vision of the cosmic landscape. Galileo's youthful fixation with mathematics had turned into an adult vocation. His contemporary and friend

Johannes Kepler was using mathematics to explore the harmonies of the universe. Like Pythagoras many centuries earlier, Kepler had a mystical bent, and was trying to spin a vision of reality that incorporated the so-called Platonic solids as the basic building blocks of the cosmos.[15] The order and regularity of the Platonic solids had triggered deep emotional feelings in Kepler, and he strove to find a way to make his inner vision of cosmic order fit the world around him.

Galileo's goals were less mystical and more straightforward. He wanted to use the tools of mathematics to study natural phenomena, to take apart and understand specific natural mysteries and problems. His friend Kepler may have wanted to weave an imaginative new vision of reality; Galileo wanted to know why small pebbles fell at the same rate as massive cannonballs.

Galileo spent eighteen years in Padua. During that time he gained a deserved reputation as a smart man with an uncontrolled tongue. He had a short temper and little patience with people who in his opinion were stupid, uninformed, or simply wrong. His penchant for savage written and verbal attacks on his associates and contemporaries in Padua got him into hot water on more than one occasion. That character flaw, combined with an inability to "play it cool" politically, would eventually get him into the biggest trouble of his life. At this point, though, Galileo was far from a standout in the crowd. Most other natural scientists and philosophers of his time engaged in the same kind of brawling intellectual discourse.

By 1597 Galileo had become convinced of the truth of the heliocentric system. He even wrote to Kepler to tell him of his conclusions on the matter. However, there wasn't much Galileo could do about it at the time. Copernicus's bottom line had been that a heliocentric system was "more elegant" than the older, more accepted Ptolemaic system of a geocentric cosmos. In fact, with the available scientific instrumentation of those times, the Ptolemaic system was actually a better fit to astronomical observations than Copernicus's heliocentric system.

More basic than this, though, in the minds of most authorities were the philosophical and theological arguments in favor of the earth-centered universe. For example, the Old Testament said quite clearly that Joshua's prayers led God to halt the movement of the Sun in the sky, so the Israelites could defeat the city of Jericho. What's more, Aristotle and the other ancient Greek philosophers (who by now were "quasi-Christians" in the minds of some Church theologians) all provided rigorous arguments about the central position of the earth in the cosmos. Not until someone could come up with indisputable evidence to the contrary would people's vision of reality begin to change.

Sometime early in 1609 a messenger passing through Padua told Galileo about the delightful new toys that were all the rage in Paris. Two lenses, placed at opposite ends of a tube, would magnify distant objects, be they ships far out to sea, approaching armies, or ladies dressing in houses down the street (there is nothing new about adolescent boys—and men—peeping through the windows of the woman down the road). Being a practical man, Galileo immediately saw the military implications of this "far-seeing tube" and set about to make one for himself. It took him about a day. His first telescope was made from a lead tube with a convex lens at the far end and a concave lens at the near end, next to the eye. It had a magnifying power of about three diameters—that is, it magnified objects to three times their actual diameter. Galileo was very pleased by the results, and soon made several other telescopes. The largest had a convex lens about 4.4 centimeters in diameter and a power of thirty-three diameters.

The "Far-Seeing Tube"

Galileo had been earning some extra money by providing technical and scientific advice to the Doge and Senate of Venice, who governed the region of Italy that included Padua. He quickly penned a letter to his political patrons, pointing out the obvious military implications for the far-seeing tube. The Doge and Senate just as quickly realized there was no way to keep the invention to themselves; it seemed everyone else in Europe was playing with the devices by then. However, they did acknowledge their indebtedness to Galileo for showing them the military use of the telescope. They increased his salary to one thousand florins per year, an unprecedented sum at the time. Galileo thus became the first "modern scientist" to earn a living as a military consultant.

The telescope that Galileo built was a refracting telescope; that is, it used lenses to focus light from distant objects and magnify them. His particular version used a very weakly magnifying convex lens as the primary lens or objective (the lens at the front of the tube) and a strongly magnifying concave lens as the eyepiece. The concave eyepiece lens sits behind the focal point of the convex lens. This combination of lenses results in a magnified image that is upright. The telescope had a relatively short focal length, so the tube holding the two lenses did not need to be very long.

However, Galileo's telescope suffered from at least two serious problems. First, the higher its magnification, the narrower its field of view. That is, the region of the sky one can observe through the telescope becomes smaller and smaller the greater one makes the magnification.

A more serious problem was poorly made lenses. Galileo had ground his own lenses, and he did a bad job of it. They had several aberrations, or defects, in the way they formed images, including chromatic aberration, coma, distortion, and spherical aberration. *Chromatic aberration* is a variation of focal length for different colors of light. The effect is a separation of light into its different colors (like a prism—about which more later), causing the image to have fringes of color to it. *Coma* is an aberration that occurs when the light from the object being observed enters the objective at an oblique angle. This causes a pointlike image such as a star to appear as a blur rather than a point. *Distortion* is an aberration that causes straight lines to look like curves. *Spherical aberration* causes the parallel light rays passing through a lens to converge not at a single focal point but at a series of focal points. With some or all of these aberrations, Galileo's telescopes did not do a very good job of imaging distant astronomical objects. That he made so many momentous discoveries is testimony to his persistence, some luck, and perhaps a little imagination.

Galileo's friend and colleague Johannes Kepler had a great mind and a brilliant way with mathematics, but he was a klutz with tools. Unable to build his own telescope, he borrowed one that Galileo had given to the Duke of Bavaria. He soon saw a way to improve its performance by using convex lenses for both the objective and the eyepiece. In this form of refracting telescope, called (not surprisingly) a Keplerian telescope, the convex lens of the eyepiece acts to further magnify the image created by the convex objective. A Keplerian refractor creates an inverted image; whatever you're looking at appears upside down. It is thus not much good for looking at distant cows, but it's perfectly fine for looking at distant stars or planets. In addition, the Keplerian refractor's field of view is larger, and its image can be magnified to a much greater power than that of a Galilean refractor. The magnifying power of a telescope is the ratio of the focal length of the objective to that of the eyepiece.[16] An astronomer can increase the magnifying power of a telescope by using an eyepiece with a shorter focal length.

The English astronomer William Gascoigne soon discovered another advantage of a Keplerian refracting telescope. By placing a thin object like a wire or two crossed wires—crosshairs—at the common focus of the Keplerian's two lenses, and then superimposing the image of a star exactly into focus on the already focused crosshairs, Gasgoigne found that it was possible to determine the position of the telescope—and the star—with considerable accuracy. Gasgoigne used this method to develop the micrometer, and by 1640 he was accurately measuring the diameters of the planets. Even today, astronomers use micrometers to measure the positions of many astronomical bodies.

New Sights in the Sky

It didn't take Galileo very long to put his *telescopio* to other than potential military use. He quickly realized that his device, which could make distant ships appear as if they were in the harbor, could also make celestial objects appear near at hand. He began pointing his primitive refracting telescope at heavenly objects and recording what he saw. What he saw was by turns well known, somewhat surprising to most people, and utterly astonishing.

Galileo began by observing the Moon, and saw that it was covered with what appeared to be mountains and oceans. It was not the perfect sphere of the philosophers and theologians. Of course, this was not exactly a "new" discovery. Any of us can tell, simply by looking at it with our unaided eyes, that the Moon is not quite "perfect." It appears somewhat mottled in color and appearance. We can see dark spots on its otherwise light gray surface. The question therefore inevitably arises: Why did the "experts" (the philosophers and theologians) continue clinging to the Aristotelian worldview, a vision of reality that proclaimed heavenly bodies to be perfect spheres, when the eye's own evidence said otherwise? The answer appears to be: Because belief is often stronger than the evidence to the contrary.[17]

In Galileo's time, most people still lived and moved within the accepted Aristotelian worldview. The imperfect nature of the Moon was well known; one could see its blemishes and bright areas with the unaided eye. No one denied it; it just didn't matter. The heavens were the abode of God; God was perfect; so all heavenly bodies must reflect the perfection of God by being perfect spheres.

Galileo's telescopic view of the Moon did not reveal something previously unknown. But his drawings of what he saw—drawings that probably had as much imagination in them as clear vision (remember, he was using a telescope with wretched lenses)—captured the popular imagination.

Galileo next turned his tiny telescope on the Milky Way, that band of light that pours across the night sky like, well, milk. Much to his surprise, he discovered that the Milky Way was made of stars—millions upon millions of them. Just as surprising was what he saw when he pointed the telescope at other areas of the night sky. He saw stars, of course, but what was important was that Galileo saw *more* stars than could be seen with the unaided eye. Obviously, the universe was considerably different from what people perceived it to be with their unaided sense of sight.

Each of these three observations—of the Moon, of the Milky Way, of the starfields of the night—was a hammer-blow on the facade of the prevailing worldview. Galileo's fourth observation was perhaps the most powerful. On the night of January 7, 1610, he pointed his telescope at Jupiter, and he saw several new "stars" near the planet. Night after night he watched the planet, and *he saw the stars move*. Galileo tracked their movements through March 2 and made drawings of them. He reached the inescapable conclusion that these new "Medicean stars," as he called them (a blatant attempt to flatter his patrons, the Medici family of Florence), were actually tiny moons circling Jupiter.

The geocentric worldview of Aristotle and Ptolemy, of the Church and just about everyone else, required that *everything* circle the earth. Clearly, everything did *not*. The Aristotelian vision of the cosmos could not be true. Copernicus's heliocentric theory, however, stated that the earth revolves around the Sun and not the other way around. Galileo's observations did not conflict with Copernicus's view. Copernicus had not asserted that *everything* revolved around the Sun, but that *almost* everything did. Those opposed to Copernicus's heliocentric theory had pointed to the "privileged" status of the Moon as a weak point. How could the Moon, they said, orbit a planet like Earth when Earth was tearing madly around the Sun? Surely the Moon would quickly be left behind. Galileo's discovery of Jupiter's moons neatly scotched that argument. Obviously heavenly bodies *could* orbit other heavenly bodies that in turn were orbiting the Sun. The Moon's status was not privileged at all. At least one other planet had moons.

If Galileo had done nothing more than write down his observations, and perhaps make a few sketches, it is possible little would have come from his first forays into telescopic astronomy. However, Galileo was nothing if not a shrewd self-promoter. He was also by now a fervent believer in the Copernican system, and he set out to convince others of the rightness of his position. In March 1610 he published a small booklet with the title *Siderius Nuncius* (*The Starry Messenger*). In it Galileo presented the results—in both words and pictures—of his first series of astronomical observations with the newly available telescope. Unlike Copernicus's *De Revolutionibus*, *Siderius Nuncius* became a best-seller, perhaps the first "popular science" book. The public was utterly captivated by Galileo's discoveries. He became famous, and he made some money, too. Orders for telescopes came flooding into his private workshop.

Galileo also garnered some criticism. Several detractors rightly pointed out that his telescopes used lenses that were often flawed or poorly ground. Some of his observations might be nothing more than artifacts caused by poor equipment. They also tried to explain away his astronomical discoveries by calling them atmospheric disturbances. As was the custom in those days, many of these attacks were of the *ad hominem* (against the person himself) variety. They were the equivalent of saying, "Galileo is a scurrilous son of a female dog in heat who lives with a woman to whom he is not married! He is a rascal, a rogue, a money-grubbing swine who toadies up to the big political figures and takes their money! Obviously, therefore, his scientific observations must be in error." That "B" did not follow from "A" made little difference. Also true to form, Galileo often struck back with equally venomous ripostes.

Meanwhile, the book sold like hotcakes and the telescope orders kept arriving. In the autumn of 1610 Galileo moved from Padua back to Pisa, and continued his astronomical observations. He examined the planet Venus and discovered that it moved through phases like the Moon, a discovery that struck another blow to the geocentric vision of reality. In the Aristotelian worldview, the Sun and the planets all revolved around Earth. But if this were true, then Venus must always lie between the Sun and Earth. From Earth, then, Venus would always appear as a crescent. In a heliocentric system, however, Earth and Venus can find themselves at times on opposite sides of the Sun. In that case, Venus will move through phases, including a "full Venus" phase analogous to a full Moon. And that's exactly what Galileo saw.

Galileo also observed dark spots on the face of the Sun. Today we know that sunspots are regions on the sun's visible surface that are cooler than the rest of the surface, which makes them appear dark. In 1610 no one knew the cause of sunspots; just the fact that they existed was astonishing enough. Galileo was not the first person to observe spots on the Sun. Humans have probably been stealing a peek at the Sun for untold thousands of years. On foggy or overcast days it is possible to look directly at the Sun for a few instants without seriously damaging one's eyes. And on such occasions it is possible to see particularly large sunspot groups on the solar disk. No doubt many had done so during the centuries before the invention of the telescope.

Probably the first person to see sunspots with a telescope was the English astronomer and mathematician Thomas Harriot. Harriot had once tutored Sir Walter Raleigh; when he was only twenty-four he had sailed to the New World as the geographer on the second expedition to what is now Virginia. By 1610 he was attached to the household of the Duke of Northumberland. Harriot was in the habit of using his own telescopes to look directly at the Sun. One can imagine how damaging to the eyes *that* practice would be. Harriot made it only slightly less dangerous by making his observations on misty mornings. Harriot was not in the habit of immediately publishing the results of his observations, however. Though indirect evidence indicates he was probably the first person to use a telescope to observe sunspots, his own earliest written accounts date to March 1612. Another early observer of sunspots was Johann Fabricius, the son of the famous natural scientist David Fabricius. Johann was the first to publish the results of his observations. At about the same time, the German Jesuit priest Christof Scheiner also saw sunspots. Scheiner performed a long and careful study of the spots and published his results under an assumed name (his religious superior did not believe sunspots could possibly exist).

Galileo, of course, always remained convinced that *he* had been the first to see sunspots with his telescopes. Years after the fact, he conveniently rewrote history when he told the story of those early days, adjusting the facts to fit his version of reality.

The Unbelievable

By July 1610 Galileo had been observing celestial objects with the newly available telescope for six months. He had already made several important discoveries: that the Milky Way was made of millions upon millions of stars; that wherever in the night sky he pointed the telescope, more stars were visible through it than with the naked eye; that the Moon was not a perfect sphere, but was mountainous and pockmarked; and that the planet Jupiter was orbited by four tiny moons.

News of these discoveries had swept across Europe. Galileo's little booklet was popular reading everywhere. That meant multitudes of people were having their inner vision of astronomical reality changed. The changes might be embraced by some, resisted by others. No matter. Anyone with a few coins could buy one of the *telescopi* available from street vendors. Those with some lens-grinding skill could make their own. Soon many others besides Galileo could see mountains on the Moon, millions of stars in the Milky Way, four stars dancing with Jupiter, and spots on the face of the supposedly perfect Sun. The new knowledge of the heavens was percolating through the popular consciousness. Within a generation it would change the culture's vision of astronomical reality.

Now the popular but controversial Italian began using his primitive telescope to observe the planet Saturn. What he saw was utterly baffling. On either side of the planet there were what looked at first to Galileo like large ears. Later he described them as two very large moons. Unlike Jupiter's moons, or Earth's own Moon, these two moons of Saturn remained stationary. Galileo reported this impossibility in a letter to his friend Johannes Kepler, and did so in the form of an anagram of thirty-seven letters:

SMAISMRMILMEPOETALEUMIBUNENVGTTAVIRAS

The anagram wasn't easy to decipher. Kepler had a go at it, thought he found the word *MARTIA*, or Mars, and tried to puzzle out a message about possible martian moons. (In fact, the two tiny moons of Mars were not discovered until 1877.) Next, some friends of Thomas Harriot tried to crack the anagram. One of them succeeded. The correct solution was the Latin phrase *Altissimum planetam tergeminum observavi*: "I have observed the furthest planet [Saturn] to be triple."[18]

However, the mystery of the triple Saturn eventually became more bizarre. The "moons" still did not move, and as Galileo continued to observe them over the following months and years, they started to get smaller. Finally, they disappeared.

This made no sense at all to Galileo, who at one point wrote, "Does Saturn devour his children?" In fact, Saturn does not have two moons nearly its own size, nor does it swallow them up. What Saturn does have is a huge retinue of very tiny moons, and some several millions of them encircle the planet as a remarkable and beautiful set of rings. What really happened when Saturn's two big "moons" disappeared from Galileo's view was that the rings were edge-on as viewed from earth. They are so narrow that they became invisible. For Galileo, though, the idea that Jupiter had four moons was radical.

But rings encircling a planet? Even for Galileo, that was a stretch. Sixty-five years later the Dutch astronomer and physicist Christiaan Huygens took another look at Saturn. Huygens's work had an enormous influence on the development of both physics (as we'll later see) and astronomy. To view Saturn, he used a telescope not much more powerful than the one Galileo used. But Huygens identified what he saw as rings. Part of Galileo's problem, of course, lay with his telescope's lenses: They weren't all that good. But another problem was one of expectation. Like everyone else, Galileo could not see something that "cannot be real." By the time Huygens was examining Saturn with a telescope, though, the worldview of Europe was already changing. Rings around a planet? Astounding! Incredible! Almost beyond belief—but not quite.

Newton's Telescope

In 1665 bubonic plague was sweeping through London. Isaac Newton left Cambridge, where he was a student, and returned to the family farm that he had inherited from his parents. Working in a darkened room in a small outbuilding, he made several major breakthroughs in our understanding of light. In doing so he laid the foundations for the modern science of optics, the branch of physics that deals with light and vision. For one thing, he discovered that when passed through a prism white light breaks into a beam of many-colored light, which he called a spectrum (from the Latin word meaning "appearance"). Newton soon proved that the colors he saw were not a product of the prism, but were a property of light itself. That is, white light was a combination of all the different colors of light. Newton also found that each color was refracted or bent in a slightly different path. This meant it was possible to determine just how a lens would refract light passing through it, and thus the degree of chromatic aberration.[19]

Newton's work with prisms led him to conclude that refracting telescopes had specific limits to their ability to magnify distant objects. He was convinced that no possibility existed of reducing, much less eliminating, chromatic aberration, a conclusion that was incorrect. In fact, various combinations of lenses made of different materials can drastically reduce chromatic aberration.

Newton didn't know this, and his erroneous conclusions led him to develop a brand new kind of telescope, one that didn't have the problems associated with refracting telescopes. He built the first working telescopes to use mirrors instead of lenses. We call them reflecting telescopes, and they are the workhorses of contemporary optical and near-infrared astronomy. The largest telescopes on Earth, the two Keck telescopes in Hawaii, are reflecting telescopes. So is the Hubble Space Telescope and all of the other new telescopes built and used today.

Newton's design for his telescope was based on the well-known ability of a mirror to reflect light that strikes it. Any reflective surface will work as a mirror: various metals, silvered glass, even the surface of water under certain circumstances. What matters is the phenomenon of reflection. Any flat mirror will reflect light rays at the same angle at which they approach the mirror. If a beam of light hits the mirror's surface at, say, a 55-degree angle from the vertical, it will reflect off the mirror at just that angle.

Now, suppose we take ten mirrors and arrange them in a half-circle with a diameter of, say, 100 centimeters. We place each mirror so that a line perpendicular to its surface goes right through the center of the half-circle, 50 centimeters distant. Now we allow parallel rays of light from some distant source of light, such as the sun, to fall upon the mirrors. The mirrors will reflect the light to the centerpoint, the focal point for all ten mirrors. Because these mirrors are flat and arranged in a half-circle, the rays of reflected light do not all come to the same precise focal point. So this "segmented mirror," as it were, suffers from spherical aberration. So does a mirror with the actual shape of a half-sphere. Light reflecting off different areas of the curved spherical surface will hit different foci.

However, suppose we take a spherical mirror and shape it slightly differently. We deepen its central area to give the mirror the concave shape of the curve called a parabola. The parabola-shaped mirror *will* focus all the light falling upon it onto a single focal point. It was this type of concave shape that Newton used in his mirrors.

Newton's reflecting telescope avoided a problem that had plagued some earlier telescopes: chromatic aberration. A Newtonian reflector has two mirrors, a concave primary mirror and a flat secondary mirror. The concave primary mirror reflects light up to the small flat secondary mirror, which sits just in front of the primary's focal point and is tilted at a 45-degree angle to the primary. Light from the primary hits the secondary mirror at a 45-degree angle and is reflected at another 45-degree angle. The light then passes through a small hole in the side of the tube and enters an eyepiece.

Simple, easy to construct, and compact, Newton's first telescopes were less then 30 centimeters long, but had the same magnifying power as a refracting telescope a meter long. And because Newton's reflecting telescope used mirrors instead of glass lenses, it didn't suffer from chromatic aberration. Its only really serious problem was the dimness of its image. Mirrors made in the seventeenth century did not reflect light with very high efficiency. Nevertheless, the reflecting telescope was destined to replace the refractor as the instrument of choice for optical astronomy. Refractors could only become so large (about a meter in diameter for the primary lens) before chromatic aberration and sheer weight took unacceptable tolls. But reflecting telescopes could—at least theoretically—be built to practically any size. The bigger the mirror, the greater the telescope's light-gathering power.

The greater the light-gathering power, the fainter the astronomical objects one can see. A faint astronomical object may be close but very dim, either emitting or reflecting little light. Or it may be something very big and very bright, and very far away. With reflecting telescopes, astronomers have been able to discover and observe both. And it was the discovery of faint objects lying at astoundingly vast distances that revolutionized our vision of astronomical reality in the twentieth century.

The Big Mirrors

The main problem with the earliest reflecting telescopes, as seen with Newton's models, was their light-gathering effectiveness. Seventeenth-century mirrors were not very efficient at reflecting the light hitting them. They were made of metal instead of glass, and had to be repolished and refigured (curved to the right shape) rather frequently. It was not until the 1850s that chemist Justus von Liebig discovered how to chemically deposit a thin film of silver on a curved glass blank, making glass mirrors feasible. Glass, of course, is much more stable chemically and dimensionally than a metal plate. Few reflecting telescopes in the seventeenth or eighteenth century could produce images as bright as those made by refracting telescopes with the same aperture. As we near the end of the twentieth century, that has changed. Various versions of reflectors have far outstripped refractors as astronomy's preeminent "light buckets"—an apt description, since large telescope mirrors can gather more light, and fainter light, than smaller mirrors the same way a large bucket can hold more water than a small bucket.

However, large reflectors also have a difficulty in common with large buckets: weight. I may be able to carry a 5-gallon bucket filled with water, but not a 65-gallon bucket. The weight of large glass lenses was a fatal limitation for refracting telescopes; now reflecting telescopes faced the same problem.

The first of the really large mirrors was the 2.5-meter diameter mirror for the Hooker Telescope atop Mount Wilson in southern California, which went into full operation in 1918. The mirror for the Hooker Telescope weighed four tons. The entire instrument—primary and secondary mirrors, steel frame, and gears to turn it—weighed 100 tons. It took several attempts to finally make the mirror. George Ellery Hale, the astronomer who arranged for his millionaire friend J. D. Hooker to finance the instrument, suffered several near nervous breakdowns during the construction of the Hooker Telescope.

Thirty years later, the Hale Telescope at the Mount Palomar Observatory in California entered its testing phase. The Hale Telescope mirror was first designed in the 1920s, and the telescope itself was finally built in the 1940s. In December 1947 the completed telescope saw first light; it went into operation on Mount Palomar in southern California in 1949. Its primary mirror, 5 meters (200 inches) across, was at that time the largest in the world. The mirror had been cast in 1934 from special low-expansion Pyrex glass, but only after going through some painful design problems. It was twice the diameter of the 2.5-meter reflecting Hooker Telescope at Mount Wilson Observatory, had four times its light-gathering power, and at 20 tons was five times heavier than the Hooker primary mirror. The cost was some $6.5 million, a large amount of money in those days. The specially curved surface of the mirror, coated with a thin layer of aluminum, had a focal ratio of f/3.3. Photographers understand what that means. For the rest of us, though, the explanation is rather simple. The focal ratio is the ratio between a telescope's focal length—the distance from the center of the curved mirror to its prime focus, the point in space where the reflected light comes to a point—and its effective diameter or aperture. "f/3.3" means that the Hale Telescope has a focal ratio of 3.3, the ratio between the telescope's effective aperture of 5 meters and its prime focus of 16.5 meters. That gives the Hale telescope a shorter tube than previous reflecting telescopes, which usually had a focal ratio of around f/5.

The Hale Telescope seemed the technological limit to the size of any telescope mirror. For any telescope much larger than that, the open steel frame tubes and supporting girders would become impossibly huge. The Hale mirror had been made in a special honeycomb pattern to reduce its weight as much as possible. Early attempts to cast it had failed before the mirror was finally successfully completed. Astronomers knew that any glass mirror blank much larger than that of the Hale Telescope would begin to sag under its own weight.

A second problem with giant reflecting telescopes has to do with a property of the very glass from which the mirror is made. Glass expands and contracts as the temperature of the air around it changes. Both the primary and secondary mirrors of reflecting telescopes have to be ground to precisely the right curve, or "figure," in order to sharply focus light onto a photographic plate or other scientific instrument. Light often acts as if it is made of waves. White light—as Isaac Newton discovered—is a mixture of light of many different colors. Different colors have different wavelengths, which is the distance from one wave crest to the next. Visible light (of all colors) comes in very tiny wavelengths, ranging from 3×10^{-7} to 3×10^{-6} meters. 3×10^{-7} meters is 0.0000003 meters, or about one hundred-millionth of a foot. The crests of all the waves in a beam of light have to stay in step and be reflected off the mirror's surface at the same angle. If the curve of the mirror is off by only one ten-thousandth of a meter, the images the telescope creates will be blurred in some fashion. It took more than a decade to grind the Hale Telescope primary mirror to the correct figure.

As long as the mirror stays still, it stays smooth. But it is part of a telescope, with an open metal tube held in a yoke of steel that tilts and twists to point the telescope at the right star. All that turning and rotating also tends to slightly warp the glass mirror.

By the first part of the twentieth century astronomers realized they would have to find new kinds of glass if they wanted telescopes with very large mirrors. The mirror of the 2.5-meter Hooker Telescope had been made of a soda-lime glass, the best available at the beginning of the twentieth century. But it was quite sensitive to temperature changes, and also took a long time to cool down to the ambient air temperature. When astronomers George Ellery Hale and Walter Adams first looked through the eyepiece of the new telescope, they were horrified to see not a crisp stellar image but rather a large blob of light. The protective dome of the observatory had been left open during the day. The curve of the four-ton mirror of the Hooker Telescope had been distorted by the modest change in air temperature inside the dome. The two paced the floor of the observatory, waiting anxiously for the mirror to cool down to the ambient temperature of the night air inside the dome. Finally, at 2:30 in the morning they pointed the new telescope at the star Vega. They saw a single bright speck: success.

Pyrex and similar borosilicate-type glasses (made from mixtures of boron and silicon) changed the way astronomers could view the universe. These "low-expansion" glasses were much less sensitive to temperature than earlier glasses. They expanded and contracted very little when the ambient air temperature changed. Pyrex and similar kinds of glass thus became the basis for the new mirrors in the twentieth century's first generation of giant telescopes.

The 5-meter Hale Telescope mirror was made of Pyrex-type glass. When it became operational in 1949 at the Mount Palomar Observatory northeast of San Diego, California, the telescope quickly proved its worth. Astronomer Walter Baade used it to tell the difference between two types of variable stars—stars that vary regularly in brightness—that were used as astronomical measuring sticks. (We'll encounter these types of stars again later on, in our exploration of science's alternate realities of cosmology.) Baade was then able to show that the universe was effectively twice as large as previously estimated.[20] Not only was this a momentous astronomical discovery, but Baade's findings also contributed profoundly to the emergence of the prevailing cosmological paradigm of the infinite, expanding universe. He could not have done it without the 5-meter mirror of the Hale Telescope, made of low-expansion glass.

However, by the early 1950s astronomers began to realize there was no way to build effective reflectors with mirrors larger than 5 meters. The technical limitations in those days were simply too great. Faced with these seemingly insuperable technical limitations, astronomers made superb use of the excellent instruments they already had. They tried not to fantasize too much about what would probably never be. Over the next three decades improved scientific instrumentation increased the seeing power of the Hale and Hooker telescopes; of the 3-meter reflector of the Lick Observatory on Mount Hamilton east of San Jose, California; of the 2.1-meter (82-inch) reflector at McDonald Observatory in Texas; and of others around the world. Advances in photographic film and developing techniques also proved a boon to ground-based optical astronomy. As we'll soon see, the development of electronic imaging technology was another trembler that would reshape our vision of the cosmos.

Astronomy began to boom as it entered the last quarter of the twentieth century and the Space Age. Even with orbiting telescopes of various kinds taking up some slack, there simply were not enough ground-based observatories on the planet to satisfy the seeing requirements of the science. Beginning in the 1970s, several new reflecting telescopes went into operation around the world. They included large reflectors at Kitt Peak, Arizona, in 1973, and at the University of Texas's McDonald Observatory. In the southern hemisphere, three large reflectors went into service in 1975 at Siding Springs Observatory in Australia and at Las Silla and Cerro Tololo in Chile.

Only one of the new telescopes had a reflector larger than the 5-meter Hale reflector—the 6-meter reflector at the Special Astrophysical Observatory in the Caucasus Mountains of the Soviet Union. And for many years that telescope was of limited usefulness. Its mirror was so large that it tended to warp slightly out of shape when the air temperature changed too much, confirming the fears of astronomers: No one, it seemed, would ever build a ground-based optical telescope with a mirror much larger than 5 meters in diameter.

Nevertheless, the new generation of large telescopes provided astronomers with additional tools to probe the heavens.

New discoveries about quasars, galaxies, and stars were made possible by the telescopes at Kitt Peak, Las Silla, Cerro Tololo, and other locations. On February 24, 1987, a new supernova, or exploding star, was discovered in the Large Magellanic Cloud, one of our Milky Way Galaxy's satellite galaxies. Dubbed Supernova 1987a, it was first seen by Canadian astronomer Ian Shelton at Cerro Tololo. Supernova 1987a was the first one seen by humans to explode in or near our own galaxy since 1604, five years before Galileo used his tiny telescope to destroy the old astronomical reality, one that was incapable of understanding or explaining supernovas.

In the late 1970s and early 1980s, several astronomers ventured forth with some radical new ideas about how to make telescope mirrors. At the same time, others proposed a new way of coping with a perennial problem faced by all users of ground-based telescopes, atmospheric distortion. Several of these ideas eventually bore fruit, breaking the "Five-Meter Barrier" for telescopes in convincing fashion. One astronomer, Jerry Nelson, was in the right place at the right time to help change our vision of reality by changing our astronomical perception of the cosmos.

A Man with a Vision

Like many other astronomers, Jerry Nelson first became interested in science as a child. When he went to college, it was to the California Institute of Technology. There he encountered Richard Feynman, taking his famous undergraduate physics classes as a freshman and sophomore. The man whose imaginative visions of electrons and quanta had revolutionized physics inspired Nelson to switch his major from math to physics. He eventually settled on astrophysics as his field.

Years later, and quite by accident, Nelson got involved in creating and building the world's first giant segmented-mirror telescope. In 1977 he was on the faculty at the University of California at Berkeley and the Lawrence Berkeley Laboratory (located atop a hill above the Berkeley campus). A committee of astronomers associated with the campus, including Nelson, had been

Figure 5. Dr. Jerry Nelson, the man who devised the Keck telescopes, stands next to Keck Headquarters atop Mauna Kea in Hawaii. (Photo by Andy Perala, W.M. Keck Observatory)

convened to review the future of astronomy at the university. Not surprisingly, one of the major topics on their agenda was telescopes. "The University of California has a three-meter telescope at Mount Hamilton [the Lick Observatory]," Nelson explains in a recent interview, "but at the time other people were building 3- and 4-meter telescopes." That meant the Lick Telescope was no longer as "big" a mirror in the astronomy community as it once had been. And that in turn meant that UC astronomers who wanted to use the biggest and best telescopes for their work would eventually be forced to find observing time on other peoples' instruments. Another problem facing the Mount Hamilton site, added Nelson, was light pollution. "The site is relatively bright because San Jose is below it. And the light pollution was getting worse. So we felt that for the University of California to stay in the forefront of astronomical research we had to seriously consider some means of augmenting our astronomical observatories."

The committee considered several possible solutions. One was building another telescope with a 3-meter mirror at a darker site. Several such locations are still to be found in the continental United States, such as Mount Graham and Apache Mountain in Arizona (both of which will be homes to new telescopes built by other organizations). However, says Nelson, "some of us thought we should just build a larger telescope. I thought it was a fascinating idea." Nelson told the committee he wanted to look at the "bigger telescope" concept in a little more detail and then report back on his findings.

"I went and studied it for a couple of months," he recalls. "Then I came back to the committee and told them, 'I think it's possible to build a 10-meter telescope, and I think it ought to be segmented.' I gave them my ideas." He chuckles and adds, "After all the initial laughter, people said, 'OK, look into it some more.' Because everyone would love to have a 10-meter telescope!" The question was, could it *really* be done? Five meters had long been the practical limit to the size of a mirror for a reflecting telescope. The Zelenchukskaya Observatory atop Mount Pastkukov in the former Soviet Union has a 6-meter mirror, but it is seriously flawed. Besides, 5 meters is not small; it's the diameter of a modest-sized living room.

Nelson's preliminary research had already convinced him, however, that a telescope mirror with twice the diameter of the Hale Telescope was feasible. He says, "I pursued it, and developed the geometry and ideas of a segmented mirror. I also detailed what I thought were the major technological issues that had to be surmounted for the idea [of a segmented mirror] to be a viable one. As I went through this process I got more and more enthusiastic about it. It really seemed that a segmented 10-meter telescope was a good idea, that it had a future, and that the problems were probably surmountable. We could build an affordable gigantic telescope."

Nelson's colleagues started catching his enthusiasm. As they became more and more convinced that a gigantic new telescope could actually be built, they got more money to do studies and prototypes. "The first funding, of course, came from the University of California and from the Lawrence Berkeley Laboratory" where Nelson worked. "From 1977 to 1985 they were the main sources of the money we used for research and development of the telescope." By 1985 Nelson and his colleagues had a mature idea of what the new telescope would be like. The University of California began looking for the large sum of money it would need to actually build it. The university's trustees knew they couldn't come up with all the money alone. They would have to either collaborate with some other institution or institutions or find a foundation with very deep pockets. That's when Nelson's undergraduate alma mater entered the picture.

Cal Tech has had a long and honorable association with astronomy, physics, and astrophysics. Many prominent astronomers and physicists (including Richard Feynman) have served on its faculty. The school runs the famous Jet Propulsion Laboratory for NASA. Cal Tech also had access to a foundation that was very interested in funding a big new telescope, the W. M. Keck Foundation. After a series of long, delicate, and at times contentious negotiations, the University of California and Cal Tech joined in a collaborative effort to actually build the observatory. The Keck Foundation would provide the dollars to do the deed. The observatory would be named after Mr. Keck, of course, following in the long tradition of naming observatories for their primary financial backers.

The Keck Observatory was conceived; its gestation period would last eight years.

The Broken Mirror

Like most of the other new telescopes built since 1985, the Keck uses a new type of mirror. In fact, the Keck mirror is thirty-six mirrors. The two Keck telescopes look out into space from atop Mauna Kea like giant insect eyes, but they see much more than any insect ever did.

The concept of a segmented mirror is in itself not new, but the ability to create an effective segmented telescope mirror is.[21] Two big advantages of a segmented mirror are cost and transportability. The cost of making a single mirror larger than 5 meters in diameter has become prohibitive. Big mirrors cost a lot of money because they contain a lot of very expensive glass and require an enormous amount of very careful, very detailed work to polish them to just the right curve. The giant mirrors of the 2.5-meter and 5-meter telescopes took astronomy to the financial limits of what could be done. A single 10-meter mirror made using the same techniques as that used for the 5-meter Hale mirror would weigh eight times as much as the Hale. The telescope mount and tube scaffolding, as well as the observatory dome, would have to be that much larger and more complex to keep such a behemoth mirror from sagging under its own weight. Then there's the matter of grinding the mirror. Counting the interruption of World War II, it took technicians eleven years to grind the 5-meter mirror blank for the Hale Telescope to just the right curve. Today, no one can afford to make even the tiniest mistake with a mirror as huge as the kind envisioned for the new generation of optical telescopes. A mistake in casting or figuring such a mirror could spell financial disaster.

One way to get a grip on the problem is to imagine that a telescope is like a car. Suppose we all drove around in cars that had only one wheel right in the middle—"unicars," like unicycles. If the tire blows and you don't carry a spare, you're stranded. If you have a spare, great. But suppose tires are incredibly expensive—something about the availability of raw materials to make tires, let's say. Tires might be so expensive that people would decide it was cheaper to get stranded, walk, or hitchhike to a pay phone, and then get your unicar towed to the nearest Tire Emporium. In real life, though, we usually don't have that problem. Cars and trucks come with four tires plus a spare. If you have a blowout, you replace the ruined tire with the spare, drive to your friendly gas station, and buy a replacement.

Astronomers who use giant reflecting telescopes are like people who drive cars with only one incredibly expensive tire. They can't afford to have a tire that's flawed, or one that will suffer a blowout. No replacements are available. Millions of people around the world learned about this state of telescopic affairs after the Space Shuttle had put the Hubble Space Telescope into orbit, and astronomers discovered its main mirror was flawed. "Hubble Hobbled" and "The Horror of Hubble" were some of the kinder headlines that grabbed people's attention.

The real problem, though, is not even the cost of specialized glass or fabrication of a mirror. The optical and engineering limits of a reflecting telescope have much more to do with light itself than with size. The mirror, no matter what its size, must be figured to an extremely fine tolerance, or curve. And it must retain that curve under often extreme conditions of temperature. It turns out that a rough relationship exists between the cost of building a traditional, large, single-mirror reflecting telescope and the size of the mirror, or its aperture. Basically, doubling the aperture increases the cost of building the telescope six times. So a 5-meter reflector like the one atop Mount Palomar costs six times as much to construct as a 2.5-meter Hooker Telescope. A 10-meter reflector using a single huge mirror and a standard framework to hold and move it (called an equatorial mount) would cost six times more than a 5-meter reflector and thirty-six times as much as a 2.5-meter. In an era of shrinking federal science budgets and tight money in general for the sciences, astronomers needed to come up with something less expensive than the "same old way" of doing things.

NASA corrected the Hubble Space Telescope's "astigmatism" with a multimillion-dollar Shuttle flight to install a special collection of mirrors and lenses. They couldn't replace the mirror itself. Nor would it be financially possible to replace a giant mirror in some ground-based telescope. However, it's a lot cheaper to replace a flawed segment of a mosaic mirror, one made of several smaller mirrors. It is basically a matter of pulling out the bad mirror segment and replacing it with the right spare.

That leads to a second advantage of mosaic mirrors. Because they are made of several smaller and relatively inexpensive segments, mosaic mirrors are easier to transport from the place where they are constructed to the observatory site. These days, that can mean a trip halfway around the world. In fact, that's just the kind of trip taken by the segments used in the Keck Telescope mirrors.

The most difficult aspect to building the Keck Telescope, says Nelson, was fabricating and polishing the mirror segments. "It was harder than we thought it would be," he recalls. "We had developed what we thought was a very clever technique, called stressed-mirror polishing. We tested it and we liked the way it worked, and we thought it was the right way to go." They then signed up the Itek Corporation to do the stressed-mirror polishing. "They had a good reputation," says Nelson, which is a bit of an understatement. Itek has been making the mirrors of many if not most of America's secret spy satellites for years now. Unfortunately, the arrangement with Itek didn't work out as they had hoped.

The team had at the beginning devised a special scheme for warping, or bending, the mirror segments to precisely the right curves by using a set of metal "harnesses." It was a fall-back position that they hoped they would never need. Ideally, said Nelson, Itek's mirror-forming and computerized polishing technology would take care of any problems with the segments' curves. When that didn't work out, however, the warping harnesses came into play. And they worked. "We deliberately adjust some springs" attached to the mirror segments, says Nelson. "These apply a fixed set of forces to the mirror segments, which deforms them to small but useful amounts."

Another piece of good fortune for the Keck team was the commercial availability of ion figuring. This technique uses a beam of ions to do the final polishing and figuring of the mirror. Ions are simply atoms that have either extra electrons or fewer than their normal complement. The result is that the atom, instead of being electrically neutral as it would be normally, carries an electrical charge. This makes it possible to direct a beam of ions using magnetic fields. Ion beams are more than just a buzzword on *Star Trek*. They're commonplace in the world of atomic and subatomic physics; physicists create and use ion beams as a normal part of doing certain experiments. What's new is their use for less experimental purposes. Doctors at some large hospitals and medical centers, for example, now use ion beams to treat certain forms of cancer. And some companies now use them to "sputter" thin layers of glass in the surface of telescope mirror segments. Says Nelson, "Ion beam figuring turned out to be a wonderful asset to us."

Adaptive Optics

Adaptive optics is the trick that really makes Keck function. Each hexagonal mirror segment sits atop a complex arrangement of pistons, pins, and disks. Some of this metal and plastic structure, pieces called whiffletrees and flex disks, absorbs strains that would otherwise twist the thin mirrors out of shape. The result is that each segment "floats" on its whiffletree-and-flex disk support as if it were weightless. These pieces also keep the mirror segments tightly locked to the telescope frame and precisely aligned with one another. The gap between each segment's edge can be no more than 3 millimeters (about a tenth of an inch) wide. Otherwise the light striking the thirty-six mirrors gets "out of step." The result would be a telescope with an effective mirror diameter of 1.8 meters—one segment.

The passive support, however, is not enough to make Keck work. This is where adaptive optics comes in. Each of the thirty-six mirror segments has three actuators attached to it. These high-precision hydraulic levers give the mirror segments gentle microscopic nudges that keep each segment exactly aligned with its neighbor. A computer controls the actuators, using data from one hundred and sixty-eight light sensors distributed along the seams of the segments. When a segment gets even slightly out of alignment with its neighbors, the sensors send electronic signals to the computer controller. The computer uses the signals to calculate exactly how out of alignment the segment is and how much or little the actuators need to move to nudge the segment back into place. The adaptive optics computer readjusts the positions of the mirrors twice each second.

Of course, actively controlled mirrors are very innovative. Nonetheless, Nelson and his associates have had no major problems with the system. "The first night we turned it on, in the telescope, it worked. It hasn't been trouble-free, of course. We have had a variety of small problems. We still have idiosyncrasies that we don't quite understand, and things that are not quite fully calibrated. But considering how unusual it was, it went very smoothly."

The W. M. Keck Observatory is now operational. The two Keck telescopes each have a focal length of 17.5 meters (57.4 feet). Each segmented mirror has an effective diameter of 10 meters, or 33 feet, making these ground-based optical telescopes the world's largest. The total light-gathering area for each mirror is 76 square meters (or 818 square feet). The thirty-six hexagonal segments are each 1.8 meters in diameter and only 74 millimeters thick—not much more than the thickness of a good-sized dictionary. Each segment weighs 880 pounds; a Keck mirror as a whole weighs only 14.4 tons. By contrast, the monolithic mirror of the Russian 6-meter telescope at the Zelenchukskaya Observatory in the Crimea, previously the world's largest, weighs a whopping 41 tons.

Keck's telescope tube, an open metal framework, is just 22 meters long and weighs 110 tons. The telescope's yoke structure holds the telescope and allows it to swivel around to aim at any point in the sky. Its Nasmyth platforms hold instruments located at the telescope's Nasmyth foci. The yoke and Nasmyth platforms are 10 meters high and have a base 12 meters square. They weigh a total of 160 tons. The total moving weight of telescope tube, yoke, and Nasmyth platforms is 270 tons. The total moving weight of the Russian 6-meter telescope is 650 tons.

Each Keck telescope sits inside a dome that is nearly 31 meters tall and 37 meters in diameter. The dome's total moving weight is 700 tons. Though it stands more than eight stories tall, the Keck telescope is very compact for the diameter of its mirror. Its squat shape gives it a remarkably short focal length; its segmented mirror is so steeply curved that it has a photographic "speed" of f/1.75.

Figure 6. The two Keck telescopes sit at the top of Hawaii's Mauna Kea, an extinct volcano, gazing out at the universe like two giant insect eyes. (Photo by Andy Perala, W.M. Keck Observatory)

The Computer-Chip Eye

Keck is one remarkably big, and remarkable, camera that allows humans to perceive the cosmos in ways they never have before. Astronomy depends on vision, but as we've seen, "vision" has been extended in remarkable ways by today's astronomers and their instruments. We can "hear" the cosmos with radio telescopes and "smell" chemicals between the stars with sophisticated instruments attached to optical telescopes. The two Keck telescopes themselves are specially designed to "see" into the infrared part of the spectrum. As we learned earlier, infrared radiation is essentially heat radiation. So Keck allows astronomers to detect—"feel," if you wish—the heat radiation from cosmic objects.

Another way to think of this process is to imagine what it would be like if you could see infrared light rather than just feel it with your skin. In fact, we have special instruments that allow us

Figure 7. This view of the Keck I telescope is from inside the dome, looking through the telescope's metal framework. (Photo © Erik Hill, W.M. Keck Observatory)

to do that, such as the night-vision goggles first used by the military and now by police.

A major advance in computer technology makes it possible for Keck and other telescopes to "see" deeper into space than ever before. Invented in 1969 by researchers at Bell Laboratories, the special computer chip called a charge-coupled device, or CCD, is designed to capture light images electronically. Like transistors and computer chips, it is made of a semiconductor. Doping the silicon of the CCD with certain kinds of elements such as germanium creates a series of positively charged regions called photosites. The photosites act as electron receptors and make the CCD photosensitive. When the particles of light called photons strike the silicon surface they knock loose electrons from silicon atoms. The electrons then jump to the receptor regions. The photosites in the specially treated computer chip are like tiny wells. The electrons knocked loose from the silicon atoms by the light hitting the chip are like drops of water falling into the tiny wells. The

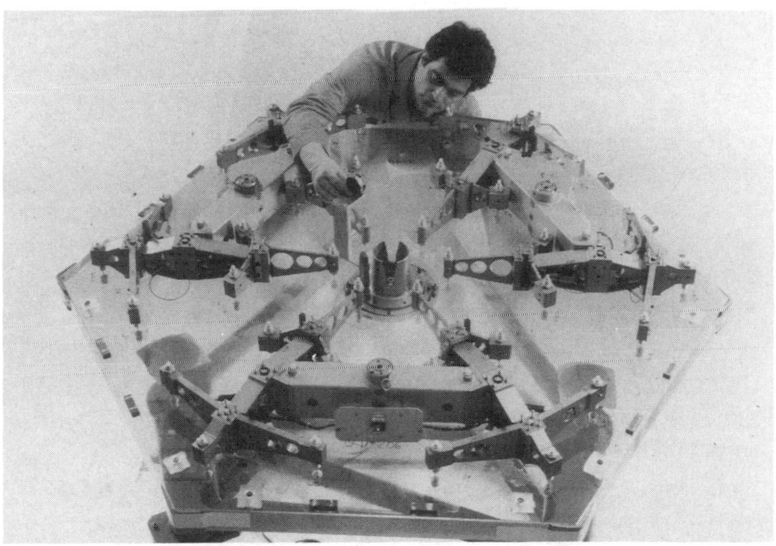

Figure 8. Each of Keck's thirty-six segmented mirrors is kept in perfect adjustment by a special arrangement of metal springs and hydraulic pistons. (Photo © Russ Underwood, W.M. Keck Observatory)

strengths of the incoming light signals are directly related to the numbers of electrons captured. The resulting electrical potentials—the electron "water drops" that accumulate in the silicon chip's tiny "wells"—are then amplified and converted to a digital code that is stored in computer memory.

Each "electron well" records the intensity of the photon striking a particular part of the CCD. The more energetic the photon, the more electrons the "well" ends up holding. Each "well" or collection of electrons is called a picture element, or pixel. The more pixels a CCD has, of course, the more detail the CCD can image.

In general, CCDs still provide less resolution than chemical-based photographic film. However, CCDs are extremely efficient light gatherers. Most photographic film has a "quantum efficiency" of 1 percent. That is, only one of every one hundred photons hitting the photographic film actually interacts with the

Figure 9. The two Keck telescopes are part of a "forest" of observatories atop Mauna Kea in Hawaii. This view shows the two Kecks in the foreground; in the background are NASA's IRTF Observatory and the CFH Observatory. (Photo by Andy Perala, W.M. Keck Observatory)

film's chemical to help create the image. Even the most sensitive film has a quantum efficiency of only 2 percent. However, many CCDs have a quantum efficiency of 60 percent. They can capture at least 60 percent of light in the visible spectrum. CCDs also have a powerful dynamic range. In other words, they can record wide extremes of light intensity within a single image.

It isn't surprising that astronomers quickly began using CCDs and associated computer programs to record images. Even many amateur astronomers use CCDs now, since CCD astronomical cameras are now available at reasonable prices for the "backyard astronomer." All of the new generation of ground-based telescopes are built to use CCDs of various kinds. Some CCDs work well with visible light. Others are designed to detect light at wavelengths the human eye cannot detect, such as infrared light. This does not mean that astronomers have abandoned the vener-

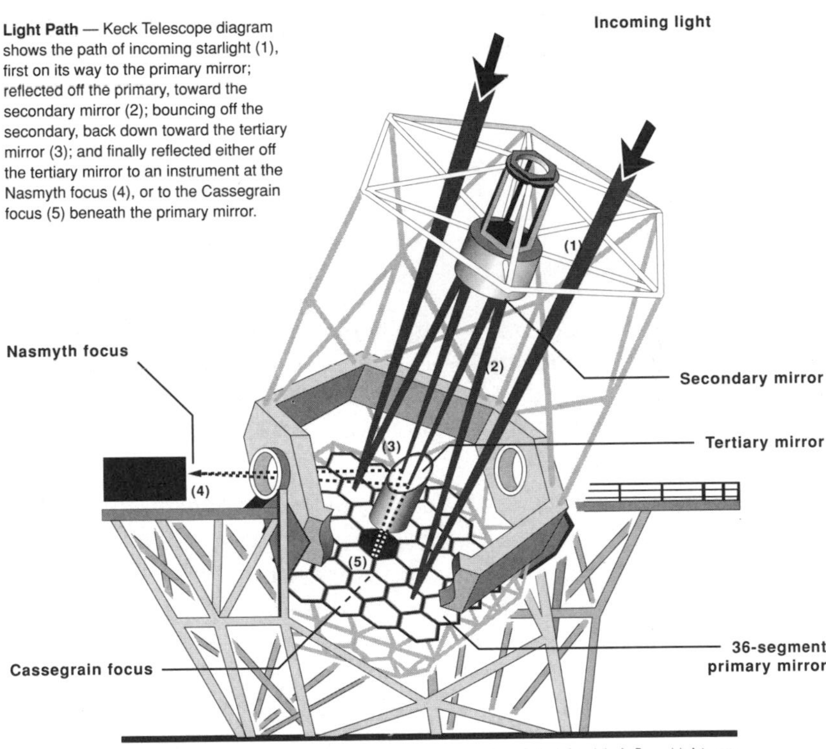

Figure 10. This illustration shows how light from distant stars and galaxies is gathered by the Keck telescope and its segmented mirror. (Illustration courtesy of W.M. Keck Observatory)

able photographic plate. Far from it. As we've seen, chemical photography still provides images of greater resolution, or detail, than most current CCDs. One of the biggest advantages of the CCD over conventional photographic film, though, is the digital nature of the information it produces. The CCD's output goes directly into a computer's memory, where the astronomer can later retrieve it and manipulate the image using sophisticated software. This marriage of the electronic imaging device and the silicon chip-based computer is one of the single most important breakthroughs in contemporary astronomy.

We can now perceive the universe in ways undreamed of less than a century ago. As we learn more about the cosmos with these new tools that expand the range of our senses, our "internal model" of the universe inexorably changes. Our inner vision of reality grows into a new shape in response to new knowledge.

HIRES

Astronomers are employing a battery of sophisticated instruments to read the information contained in the light striking the Keck's mosaic eyes. One is HIRES—"High-Resolution Echelle Spectrometer." It took about five years and $4 million to design, build, and commission HIRES.

The spectroscope, spectrograph, and spectrometer are all instruments that create, examine, or record spectra. A prism is the simplest form of a spectroscope. Most spectrographs and spectroscopes today use diffraction gratings instead of prisms to create a spectra. A diffraction grating is a series of equidistant parallel lines (usually about seventy-five hundred per centimeter) ruled on a piece of metal or glass with a fine diamond point. The closely spaced lines bend, or diffract, light (thus the name) of different wavelengths at slightly different angles. This breaks light up into its constituent colors. By controlling the shape and size of the grooves when making a grating and by shining a light on the grating at different angles, it is possible to create spectra with a purity and brightness far greater than any produced by a glass prism.

The man hired to create and run HIRES is Dr. Steve Vogt, the lead scientist for the instrument. Resolution, says Vogt, has to do with how finely a beam of light is broken up into distinct colors. "Formally, it is the ratio of the wavelength of light divided by the smallest wavelength interval between distinguishable colors," Vogt explains. "Typically, high resolution means this ratio is twenty thousand to one hundred thousand. Low resolution means the ratio is less than about five thousand." According to Vogt, astronomers often trade off resolution for faintness. That is, they

Figure 11. Steve Vogt is in charge of HIRES, a special scientific instrument used at the Keck Observatory that deciphers the messages contained in the light from stars and galaxies. (Photo courtesy of UCO/Lick Observatory)

may sometimes give up spectral resolution in order to see fainter objects. HIRES, however, has been built to do both, to see very faint objects in the universe and also have a high spectral resolution. It is astronomy's attempt to have its cake and eat it, too.

Says Vogt, "A high-resolution spectrometer breaks light up into more channels. It does this with an echelle grating, which is a kind of diffraction grating." The result is that HIRES breaks up the visible region of the spectrum into more than 30,000 distinct colors.

Vogt has been reading the messages in starlight since the mid-1970s, when he was a graduate student at the University of Texas McDonald Observatory. "I worked mostly on stars, studying star spots and stellar magnetic fields. When I came to Lick Observatory I continued to do that."

When the University of California wanted to build a 10-meter telescope, Vogt and other Lick Observatory astronomers offered some radical instrument proposals. They wanted Lick to stay a major player in this project. "Joe Miller proposed a low-resolution spectrometer," Vogt recalls. "I championed a high-resolution spectrometer." In the end those two instruments were picked as the optical instruments. However, by then Cal Tech was in the game, and politics and courtesy required that they get one of the optical instruments for Keck, so Miller and his instrument moved to Pasadena.

The high-resolution spectrometer's work begins at the so-called Nasmyth focus of the Keck telescope. The telescope's light is brought to a pinpoint focus and then passed through the narrow, precise jaws of HIRES's entrance slit. The light beam expands after it goes through the focal point. Immediately beyond the slit, the beam passes through several "photon sorters," which are simple colored-glass filters for rejecting photons of unwanted colors. It then hits a special spherical mirror, which collimates the expanding light into a beam 30 centimeters (nearly a foot) in diameter—a good-sized beam. "We use two collimators," Vogt explains. "One is a blue-optimized mirror. That is, it is specially treated for high reflectance of blue light. The other is a red-optimized mirror. It's coated with a reflective surface we call the 'Holy Grail,' a special silver recipe with sapphire overcoating. It's the 'silver chalice,' as it were."

Next, says Vogt, the collimated beam is directed onto the echelle grating, a special kind of diffraction grating used in spectrometers that does the fine-sorting of photons by color. The high resolution of the spectrometer results from the use of this very large echelle grating. Notes Vogt, "We use an echelle grating that is tipped at a very high angle to the light beam, about 75 degrees. The light beam hits the grating so steeply that the footprint of the beam is an ellipse now 48 inches across. This means we need a big echelle, much bigger than the previously largest gratings ever made." These have been about 8 by 12 inches, the size of a sheet of notebook paper. "We had a 12- by 16-inch echelle grating fabricated, and then had three identical copies made and assembled into a single large mosaic grating. They all have absolutely the same shape. They are made of zerodur glass and mosaicked together, which is a tricky engineering process. Other grating mosaics have been made before, but none had been stable in a totally passive sense."

Next, Vogt and his colleagues turned to an unlikely subcontractor: a tombstone cutter. "He made us a big piece of granite," explains Vogt, "and we attached the gratings to the granite with a carefully designed support system." The granite block provides the mosaicked echelle gratings with the stable backing essential to keep them aligned. Vogt and his team also had to take into account the flexing of the granite itself at the atomic scale. By doing so they were able to eliminate all the tilting in the gratings and all thermally induced warpings. Each echelle of the mosaic sits on three quartz washers, polished to just the proper thickness until the pitch and yaw of each was exactly the same and the assembly was perfectly aligned.

After hitting the echelle grating mosaic, the light beam passes to a cross-disperser grating. Explains Vogt, "This is a low-dispersion grating that sorts the light from the echelle into short strips or bars that fall on a detector. Each bar of light is a piece of the spectrum, with each end of a particular bar joining the other end of the bar above it and below it." So the spectrum gets sliced up and packed into a rectangular format that fits the shape of the square detectors.

"Once the beam diffracts off the echelle and cross-disperser gratings we've got photons coming out at different angles, and we've got to get them down to a focal plane really fast, in a very short distance," says Vogt. "That's because the telescope is big and the detector is small. So we have to compress light down to a small focal plane. That means we need an extremely 'fast' camera. Most people think that f/1.4 or f/1.8 is fast for a camera lens. In fact, what we need is an f/1.0 camera, a screaming fast camera with a mouth 30 inches in diameter and focal length of only 30 inches."

Vogt and his colleagues wanted to see light as blue as 3,000 angstroms. This is light in the ultraviolet part of the spectrum, far bluer than the unaided eye can detect. But most glass is opaque to ultraviolet light. Only a few types of glass can transmit ultraviolet light, like calcium fluoride and fused silica. However, calcium fluoride is like salt; it absorbs water easily and cannot yet be made in very large sizes. So Vogt's group had to use fused silica glass.

On the other hand, the HIRES team also wanted to see up into the red part of the spectrum to around 10,000 angstroms, a huge chromatic range. And they wanted to do all this without having to refocus the camera. "Most cameras work in the range of 4,000 to 7,000 angstroms," says Vogt. "Just three to four years ago we thought it was impossible to do what we wanted. But after several years of intensive computer-aided design work, we pulled it off."

The key to this success was HIRES's optical designer. Harland Epps is a professor of astronomy who used to be at UCLA and is now at Lick Observatory. Over the last twenty years Epps has developed computer codes that help design telescopes and all manner of extremely sophisticated optical systems. "He's worked with people in Arizona, Japan, Harvard, and others all over the world in optical designs for instrumentation for astronomy," says Vogt. "I worked closely with him on the HIRES camera design. He said it was impossible to design a camera of this kind. But we kept pushing, and pushing. And we finally began to see a new solution. After three years of design analysis and computer coding Epps finally came up with a camera that's absolutely spectacular!"

So now, says Vogt, "we have an f/1.0 camera that's totally achromatic. The lenses are made of fused silica with all spherical systems. It works with all colors of photons from 0.3 to 2 microns and delivers these photons precisely to a flat focal plane and with no refocusing for the different colors. This is the real miracle of the high-resolution spectrometer."

The two lenses of the camera in HIRES are so big that they sag under their own weight. Vogt's team had to build a support system to bend them back into the right shape. Each lens is about a meter in diameter. One is about 5 to 6 inches thick in the middle, and the other is about 4 inches at the edge and just 1 inch at center.

The light passes through the two lenses, and then to a mirror about 45 inches in diameter with a curvature of f/0.76. "It looks like a solid glass satellite dish!" Vogt says with a laugh. "The mirror is very light; it has a honeycomb structure and is made mostly of air. It weighs 180 pounds instead of half a ton." HIRES's camera is like a 1.2-meter telescope lying on its side: a telescope inside a telescope.

Vogt continues: "The light now comes to a prime focus, into a dewar bottle and then to a special detector." A dewar bottle is a special kind of container that holds a liquid gas like nitrogen or helium. Astronomers and physicists use dewar bottles to cool down instruments that can function effectively only at very cold temperatures. In this case the dewar cools down a very large CCD—which, as we noted earlier, is a highly sophisticated electronic "camera." The CCD used in HIRES is 50 millimeters on a side and has four million pixels, or picture elements. The CCD is the end of the line for the photons that have come through HIRES. There they are finally captured, their energies read out, and the information passed on to a computer.

"The CCD in HIRES is exquisitely sensitive," explains Vogt. "It has about 80 percent quantum efficiency. It does have a few bad pixels, because it's an engineering-grade device rather than a science-grade detector. But it's an engineering-grade device that most astronomers would love to have. In fact, the CCD has a big felt-tip pen mark on it because it was considered engineering grade, not science grade. The engineers at Tektronix, the company that built it, apparently are instructed to mark the CCDs that don't pass muster with a felt-tip pen. But we still can do lots of excellent science with this device. And that pen mark is no doubt the best-studied felt-tip pen mark in the world!"

As the photons come into the instrument, explains Vogt, "we have to figure out how many are at each wavelength. Between the coarse sorting by the filters, the very fine sorting by the echelle, and the moderately fine sorting by the cross-disperser, each photon's exact color is revealed by where it falls on our detector.

Knowing the photon's location on the detector thus gives us its unique wavelength. The detector doesn't count each individual photon. It counts about five photons at a time. A number of photons will build up at one place; then the detector measures the intensity at each location. That gives the number of photons at each and every color of the spectrum."

Why go to all this trouble to capture a few particles of light? Vogt explains that astronomers must use "big telescopes to do high-resolution spectroscopy because once you get the photons so finely sorted by color, there's very little light left in any one channel. The signal gets real weak, and so you need a telescope that gathers lots of photons. You want to observe objects that are very faint because these are the objects that are not well studied. The brighter objects in the sky have all been studied for a long time. And you want to push your high-resolution studies to as faint as you can go. So we'll be looking at quasars at eighteenth and nineteenth magnitude. We also want to look at the brightest stars in other galaxies. That means looking at stars that are ten to twenty million light-years away and measuring their velocities in other galaxies. We will be able to use that information to measure directly the presence of dark matter in other systems."

Adds Vogt, "But the big thing that will come out of high-resolution spectroscopy over the next four to five years will be the absorption spectra of extremely distant quasars. With this new instrument we can work ten to a hundred times faster than what has ever been done before with other spectrometers on other telescopes. In fifteen minutes you can do what took other people three days."

Visions of Distant Space and Time

A telescope is a kind of time machine. Light travels at a constant finite velocity of nearly 300,000 kilometers per second. That's about 1.08 billion kilometers (670 million miles) per hour, or about 9.47 trillion kilometers (5.9 trillion miles) in a year—one light-year of distance. Light takes time to get from one place to another. When we look at the Moon, therefore, we do not see it as it is at the moment, but rather as it was about one and a half seconds ago. The Sun we see in the sky is not the Sun as it is "now" but as it was about eight minutes ago. If the Sun suddenly exploded "at this moment," we wouldn't know it for another eight minutes.

Light takes several hours to reach us from the solar system's outer planets of Uranus, Pluto, and Neptune. And the stars are much farther away than the farthest planets. Even Proxima Centauri, the star nearest to our solar system, is so distant that its light takes more than four years to reach us.

The Keck Observatory's two telescopes are the world's most powerful ground-based "time machines." Their enormous 10-meter mirrors gather light from the most distant objects in the universe. The quasars and distant galaxies it espies lie nearly 15 billion light-years away. We look deep into space with Keck, and we also look deep into time. Fifteen billion years ago, the universe itself was still quite young. Keck will enable astronomers to trace the evolution of galaxies over billions of years of time. Keck may even give us views of galaxies in the very act of forming, condensing from primordial clouds of hydrogen and helium gas formed only a few hundred thousand years after the universe itself appeared.

Keck will also give us new images of stars and other astronomical objects closer to home. Faint stars at the outermost edges of our own Galaxy, the Milky Way, will reveal new information to us under Keck's gaze. These stars date from the formation of the Milky Way itself billions of years ago. We will learn more about how our home galaxy came into being.

One of Keck's strong points is its ability to "see" in the infrared region of the spectrum. Infrared light gives us a totally different view of the cosmic landscape spread out across the night sky. Keck in particular will be able to see through the dense clouds of gas and dust that obscure parts of the Milky Way. Unlike visible light, which cannot penetrate these interstellar clouds, infrared radiation passes through fairly easily. Many of these interstellar clouds are the birthplaces of new stars. Keck will be able to spot new stars forming, and perhaps even planets circling these and other stars in our Galaxy.

We have already indirectly detected the presence of giant planets circling distant suns. But to actually see planets circling other stars will utterly change our vision of reality. We will unambiguously know that our solar system is not a fluke, not unique. We will know that there are other continents out there, on the other side of the ocean of space. Telescopes have already expanded our vision of the cosmos far beyond the solar system. To actually see "with our own eyes" a new planet around another sun is a kind of knowledge that is much more "real" and closer to home than knowing the universe is 15 billion light-years wide.

Even if Keck does not spot extrasolar planets, it will surely reveal to our vision new cosmic wonders whose existence we are utterly incapable of predicting. Keck may even put us in Galileo's position: looking at something analogous to Saturn's rings and being completely unable to understand what we see.

Galileo's discoveries of four moons of Jupiter, sunspots, and the true nature of the Milky Way did more than revolutionize astronomy. Those discoveries and others that followed changed Western civilization's visionary model of the cosmos. Galileo and his successors in the seventeenth and eighteenth centuries forged an alternate vision of reality. In the 1920s and 1940s, astronomers like Walter Baade and Edwin Hubble (whom we will soon meet) made astonishing discoveries about the vastness of the astronomical universe using the then-new 2.5-meter and 5-meter telescopes of the Mount Wilson and Mount Palomar Observatories. Those observations, too, have contributed to a new picture of the cosmos, another alternative astronomical reality. Like Margaret in Wordsworth's poem *The Two-Part Prelude*, the telescopic and inner eyes of the astronomers have been "busy in the distance, shaping things / Which [make their hearts] beat quick." Also like Margaret, our eyes and the eyes of astronomers have not detached themselves from our bodies "in order to travel to the far horizon," as Wordsworth scholar Duncan Wu wrote. But our minds, individually and in concert with others in our culture, have shaped what astronomers have seen in the far distance into an alternate reality.

Keck and its sister instruments, like that absurdly tiny *telescopio* used by Galileo more than three hundred years ago, will continue to shape and change those visions by allowing us to see new and wondrous things invisible.

And one of the alternate realities these metaphorical "eyes" and "ears" have created and changed is the biggest vision of all: Western civilization's cosmological paradigm.

3

Ripples of Light

There was a time when meadow, grove and stream,
The earth and every common sight,
To me did seem
Apparelled in celestial light,...

—William Wordsworth
Ode: Intimations of Immortality from
Recollections of Early Childhood

Edwin Hubble was born on November 20, 1889, in Marshfield, Missouri. Pictures of him both as a young man in his early twenties and as an elderly, honored astronomer reveal that rangy, midwestern look. Hubble was tall and lanky, with a plain, all-American face and an air of quiet confidence. While still a child he and his family moved to Chicago. There he attended high school and received a scholarship to the University of Chicago. Hubble studied mathematics and astronomy, inspired and encouraged in part by Robert Millikan and George Ellery Hale. A brash young astronomer and teacher, Hale was the driving force behind the construction of the Hooker and Hale telescopes in California. Millikan was a physicist and educator who taught at the University of Chicago from 1896 to 1921 and later at the California Institute of Technology in Pasadena, where he was director of Cal Tech's Norman Bridge Physics Laboratory. In 1923 Millikan received the Nobel Prize in physics for his measurement of the charge of the electron and his exploration of the photoelectric effect. Millikan was mentor and inspiration to a whole generation of physicists and astronomers, and Hubble must surely rank as one of the most famous of his students.

After receiving a B.S. in astronomy and mathematics in 1910, he was named a Rhodes scholar and headed off to England and Oxford University. Hubble was also interested in law—his father was a lawyer—and it was jurisprudence (and Spanish) that he

Figure 12. Physicist Robert Millikan was a mentor to many other distinguished scientists, including Edwin Hubble.

studied for three years at Oxford. In 1913 Hubble returned to the United States and taught for a year in New Albany, Indiana.

But astronomy continued to call him, and he could not ignore it. In 1914 he enrolled as a graduate student in astronomy at the University of Chicago, working at the Yerkes Observatory near Lake Geneva, some 130 kilometers northwest of Chicago. (The Yerkes refractor, with its 1-meter-diameter primary lens, is still the largest refracting telescope in the world.) That year Hubble heard the announcement of an astronomical discovery that helped set his course. The discovery, made by another astronomer born in the American Midwest, was based on information carried on waves of light. It had to do with the astronomical objects known as spiral nebulae. Ten years later Hubble would be living and working some 2,000 miles to the southwest of Yerkes, in "Earthquake Country." There he would observe the cosmos with the largest reflecting telescope on earth, built through the efforts of his college professor George Hale. Hubble's work would change the face of cosmology the way the 1989 Loma Prieta earthquake changed the landscape of Santa Cruz, California.

The Science of Beginnings

Cosmology is the scientific study of the origin of the universe, its chemical elements and large-scale structure. Not all scientists agree that cosmology is a science, however. "We are often looked down upon by people in other disciplines," wryly notes cosmologist George Smoot. With his brown hair and neatly trimmed, graying beard, Smoot looks a bit like a caricature of an English professor. But he is one of the people working at the expanding edge of scientific discovery, in the great tradition of Edwin Hubble. The reason for the skepticism from other fields, Smoot adds, is pretty straightforward. Until just a few decades ago cosmology was a science built upon a wealth of speculation, with practically no data beneath it.

Speculation—even of the informed variety—is fun, but it is not science. Science is built upon the accumulation of repeatable observations and experiments. Observations and (where possible) experiments yield facts, information. Scientists then create theories to explain the observed data and use the theories to make predictions about the existence of unseen facts or phenomena. Then they observe and experiment again. And again. And again. If the predictions prove true, then the theory may well be a reasonably valid description of reality. If the predictions don't pan out, then the scientists (eventually, often reluctantly, perhaps even kicking and screaming) go back to the drawing board.

Biology, botany, chemistry, genetics, petrology, physics—these are sciences that lend themselves to experimentation. Doing science with rocks, or small mammals, or chemicals in beakers, or flowering plants is relatively straightforward. The object of study is close at hand. Scientists can easily carry out repeatable experiments and observe the results of those experiments with little difficulty. The instruments they use may get bigger, or smaller, or more expensive, but the science itself is fairly simple to carry out.

This is not so with the so-called observational sciences such as astronomy, astrophysics and some other branches of physics, evolutionary biology, areas of geology, and cosmology. In these cases, the objects of study may be spread out across vast areas of time or space. In the case of astronomy, the object studied is the macrocosm itself. One cannot carry out repeatable experiments on a supergiant star that's more than 500 light-years from Earth. In the case of cosmology, it's the whole ball of wax—space, time, matter, energy, and how it all got started in the first place. One cannot create variations on a universe in one's laboratory and see what develops in fifteen billion years.

Until the mid-1920s, in fact, cosmology as a science had very little in the way of hard facts with which to work. Once this situation began to change, though, it changed with a vengeance, until cosmology, the science, profoundly changed the way we perceive reality. Several astronomers and cosmologists played an important role in engineering this shift in our perception of reality. Edwin Hubble, though, stands head and shoulders above the rest. He revolutionized our view of reality by discovering previously unknown messages hidden in the light from distant stars.

Astronomy and Early Scientific Cosmology

Many early users of telescopes had spied objects in the skies that were not sharp bright points of light like stars, but rather faint and fuzzy in appearance. As early as 1661 (only fifty years after Galileo first pointed his primitive telescope at the moon) telescope-users were calling these objects nebulae, from the Greek word *nephos*, or "cloud." At least two nebulae are visible to the naked eye. One in the constellation Andromeda, the Andromeda

Nebula, has an elongated shape that is easily visible to the unaided eye. The Orion Nebula, a diffuse patch of light in the sword of Orion, is just barely visible to the naked eye. The first written references to both appear just a few years after the invention of the telescope—the Andromeda Nebula in 1614 and the Orion Nebula in 1619. At the end of the seventeenth century Christiaan Huygens, the first person to comprehend what Galileo could not—the existence of rings around Saturn—drew a picture of the Orion Nebula in his diary.

As telescopes got better, observers found more and more nebulae. Some turned out to be made of stars; others appeared fuzzy and ill-defined in even the best telescopes. Mostly, though, nebulae were a figurative pain in astronomers' necks because they kept being confused with comets. Many an astronomer in the first decades of the eighteenth century announced the discovery of a new comet, only to realize he had simply found another damned nebula. In 1771 the French comet hunter Charles Messier decided to do something about it. He compiled a list of astronomical objects that were not stars and that were frequently confused with comets. Messier was not too concerned with what these nebulae really were, only with what they were *not*. They were not comets. Messier's catalogue included one hundred and three different objects in its 1784 edition. Six more were added in 1786. Each object was numbered, with the number preceded by the letter M, for Messier. Even today, many of the objects first listed in the catalogue are frequently referred to by their Messier number. The Andromeda Galaxy (or Nebula) is still often referred to as M31, for example, and the Orion Nebula as M42.

Other astronomers also compiled lists of nebulae. The most notable catalogues were those of William Herschel and his son John. By the end of the nineteenth century the two of them had identified more than five thousand nebulae. John Herschel published the *General Catalogue of Nebulae* in 1864. The *New General Catalogue*, published in 1908, and its supplement, the *Index Catalogues*, listed more than fifteen thousand nebulae.

What were these nebulae? No one really knew for sure. Viewed with strong enough telescopes, some turned out to be clusters of stars. M13, for example, is a globular cluster also known as the Great Cluster in Hercules. Today we know that globular clusters are great spherical collections of stars that orbit a galaxy in immensely elongated orbits. Edmond Halley of comet fame originally discovered the Great Cluster in 1714. Others, however, stubbornly resisted any resolution into stellar clusters. They remained fuzzy and elusive. William Herschel, for one, suspected that some nebulae might be huge collections of stars lying outside our own Galaxy. He had no way to prove his suspicion, though, and did not pursue it.

Some fifty-six years after Messier had compiled his list of noncomets, an English lord made a major discovery about some nebulae. William Parsons had been born in 1800 in York, England, the eldest son of the second Earl of Rosse. Parsons first went into politics, eventually taking a seat in the House of Commons for King's County in Ireland, the site of his family's castle. Despite (or perhaps because of?) his reputation as one of the few honest men in Irish politics, Parsons felt little enthusiasm for his profession. What he was enthusiastic about was telescopes. Big telescopes.

Parsons began by building reflecting telescopes with mirrors 15, 24, and 36 inches in diameter. Parsons's success led him to resign from Parliament in 1834 to devote himself full-time to astronomy. In 1841 his father died and he became the third Earl of Rosse. Now he was free to use all the facilities of his Irish estate for his hobby. His ultimate achievement was a telescope dubbed "The Leviathan of Parsonstown." Completed in 1845, the telescope had a mirror 180 centimeters (72 inches) in diameter, weighing four tons. Made of speculum, a metal alloy of copper and tin, the mirror was very accurate for its time. Although the telescope itself was far from perfect, it was in many ways a triumph of engineering and design.

Lord Rosse quickly proved how valuable the Leviathan would be to astronomy. Shortly after its completion, he used the telescope to observe the nebula listed in Messier's catalogue as M51. To Lord Rosse's surprise and delight, the nebula had the shape of a spiral. Today we know M51 as the Whirlpool Galaxy; Lord Rosse's 1845 drawing of M51 is an accurate depiction of the galaxy's spiral structure.

Neither the Leviathan nor other telescopes of that time could discern any stars in M51. It was clear, though, that this nebula had a spiral shape. As the eighteenth century progressed, so did the quality of telescopes that followed the Leviathan. Other nebulae also proved to have a spiral structure when viewed through the eyepieces of powerful telescopes, though no individual stars could be seen. Most nebulae, however, began to surrender their identities as star clusters to the powerful light-gathering ability and resolution of the Leviathan and its successors in Europe and America. Lord Rosse himself, the discover of the first spiral nebula (M51), was convinced that even the brightest of them all, the Orion Nebula, would eventually prove to be a cluster of stars. Thus the "star cluster nature" of nebulae became a small piece of reality for astronomers. They had come to trust utterly in the infallibility of their technology, the total objectivity of their lenses and mirrors, the truth of this little part of their larger worldview.

But once again, this "truth" would turn out to be only a fragment of something larger. And again, it was a change in the mode of human visual perception that opened a new window onto reality.

New Visions of Light

In 1830 the philosopher Auguste Comte published the first portions of his monumental philosophical work *Cours de Philosophie Positive*. Comte is today referred to as the father of positivism. This system of thought declared that the goal of knowledge is simply to verify the phenomena humans experience—and nothing more than this. Comte's positivist philosophy heavily influenced much of nineteenth- and twentieth-century thought and philosophy. In particular, the offshoot known as logical positivism has had a profound effect on the philosophical underpinnings of twentieth-century science.

In 1835, in one of the volumes of his great work, Comte tried to explain the concept of the absolutely impossible. As an example, he offered the idea of anyone ever being able to determine the chemical composition of the stars—this, Comte declared, was "eternally impossible." What Comte didn't figure on, of course, was that humans would find new ways to decipher the messages carried in the light from stars. Isaac Newton had laid the foundation for this shift in our visual perception of the cosmos back in 1665.

Newton had been a frail infant, and his father had died soon after Isaac was born. When he was only two, his mother abandoned him to the care of his maternal grandmother and remarried. When he was in his teens he briefly showed some romantic interest in a neighbor girl. However, Newton dropped his half-hearted pursuit of her because—he is supposed to have later said—such an involvement would have distracted him from his intellectual pursuits. He never formed a romantic attachment to another woman. He is known to have suffered panic attacks at the prospect of his work being published. He suffered at least two nervous breakdowns and exhibited various forms of bizarre behavior, some of which may have been caused by mercury poisoning from his alchemical experiments. He was fascinated by alchemy. Newton was also mean, vicious, and vindictive when he felt his ideas or theories were challenged. The most famous example of his mean streak was his vendetta against the German mathematician Gottfried Leibniz. Leibniz and Newton both independently invented calculus, the powerful form of mathematics that allows scientists to study continually changing quantities. Newton was incensed by the thought that he had to share the glory with someone else. He had his friends and supporters spread a series of vicious rumors about Leibniz throughout England and Europe. That Leibniz had somehow stolen the idea for the calculus from Newton was only the least poisonous of them.

For all his emotional and psychological frailties, however, Newton was an undoubted intellectual giant. After trying and failing to run his late father's farm in Lincolnshire, an uncle connected with Cambridge University's Trinity College got Newton into the university. He had just received his degree in 1665—after a rather undistinguished career as a student, it seems—when the Black Death swept through London. It was then that he retreated back to Lincolnshire and the farm to continue his thoughts, writings, and experiments. In a small outbuilding on the farm Newton discovered that white light is a plain box containing enormous treasure. Newton opened the box. His key was a triangular-shaped piece of glass called a prism.

The story goes that Newton, while sitting in the darkened room of the outbuilding, noticed that a small hole in the wall had let in a beam of sunlight. The light beam made a small spot of white light on the back wall. When he placed a prism between the light beam and the back wall, Newton discovered something wonderful: A stream of colors poured out of the prism and bathed the wall. Newton saw that white light from the Sun is actually made of many different colors of light. This visible spectrum, as Newton named it, includes the colors red, orange, yellow, green, blue, indigo, and violet. Nearly everyone has seen one kind of naturally produced spectrum: a rainbow. (Newton first used the word spectrum in 1671, in a paper about his experiments printed in the *Philosophical Transactions* of the Royal Academy.)

Newton's discovery of the spectrum was only one of many findings he made about the nature of light. He engaged in a long series of experiments and speculations, and as he eventually wrote in his book *Optiks*—not published until 1704—his experiments led him to believe that light was made of minute particles. He called them "corpuscles" (the word first appeared in English in 1660 and comes from the Latin word *corpus*, or "body"). Today we call them photons. According to Newton, when light corpuscles of different colors passed through a glass prism, the different corpuscles were bent at varying degrees toward "the normal" by their varying attraction to the denser glass of the prism.

Newton's corpuscular theory of light was not the only one making the rounds of Europe in the late seventeenth century. Another scientist had come up with a competing explanation. In 1678 Christiaan Huygens (pronounced "HI-gens" with a hard "g") announced that his experiments had led him to conclude that light acted like a wave. According to Huygens, light consisted of wave-like motions passing through a medium that permeated all of space. Huygens, along with nearly everyone else both then and today, could not conceive of waves without a medium to carry them. The evidence of one's senses provided ample proof. Consider (Huygens and others might have said) the two most common forms of wave phenomena we encounter in our daily lives: sound and ocean waves. Sound consists of pressure waves in air, which is the medium. Waves in the ocean or in lakes are movements in water. How can one possibly have ocean waves without water? Or sound without air? If one removes the water, one eliminates the wave. Take away the air, and sound waves disappear.

The medium for light waves was called the ether (sometimes spelled aether), a philosophical descendant of the "quintessence," the ancient Greeks' fifth basic element. The ether possessed no physical properties at all; it was colorless, odorless, weightless, possessed of no mass or energy of its own, and utterly undetectable. It simply existed, and in doing so provided the medium through which waves of light could pass.

Different colors have different wavelengths, explained Huygens, and the different wavelengths travel at differing velocities as they move through the glass of the prism. When they exit the prism they have been separated from one another and so reveal themselves to the eye. Voila: the rainbow colors of the spectrum.

Huygens and Newton thus had profoundly different views of the nature of light, space, and the cosmos. Newton believed space was essentially empty and that light traveled through it at an extremely rapid velocity in the form of corpuscles. Huygens, by contrast, perceived of the space between the planets, the Sun, and the stars as filled with the undetectable ether. In fact, for Huygens and his supporters, the existence of ether also provided a concrete explanation for the action of Newton's other earth-shaking discovery: gravitation.

A New Vision: Gravity

Newton was not the only person of his time to wonder about the force of attraction that held the solar system together. Huygens, for example, had also explored one part of the question in his book *Horologium Oscillatorium* (*On Pendulum Clocks*). Huygens had invented the pendulum clock, and in this book he examined some of the mechanical and mathematical problems associated with pendulums and the movements of bodies. The English scientist Robert Hooke may also have done some work on gravity. Hooke was the first scientist to pose a theory of planetary movements as a mechanical problem. After Newton published his findings on gravitation, Hooke asserted that *he* had known all about that stuff for a long time; indeed, Hooke added, Newton may have stolen the idea of gravity from him. It is more likely, though, that Hooke's conversations with Newton merely helped draw Newton's attention to the problem. Others who at least talked about the attractive force between the Sun and the planets included Christopher Wren, the great English architect, astronomer, and mathematician, and Edmond Halley, the astronomer for whom Halley's comet is named and who was probably Newton's closest (if not only) friend. None of them, however, could come up with a rigorous mechanical and mathematical explanation for the attractive force. That fell to Newton.

As with his work on light and calculus, Newton had also begun his exploration of gravity on the farm at Lincolnshire. The well-known apocryphal story (first mentioned by the French philosopher Voltaire) has Newton observing an apple falling from a tree and wondering: What caused that? He then went on to wonder just how high up the apple could be and still fall to the earth under the influence of this mysterious force. Could this attractive force between Earth and apple extend as far as, say, the Moon? If this tale is true, Newton was engaging in what Albert Einstein would later call a *gedanken experiment*—a thought experiment. If the attractive force did extend as far as the Moon, then perhaps the Moon was falling *around* the earth. Bells started going off in Newton's head. He was on the hunt.

By 1618 Johannes Kepler had formulated his three laws of planetary motion in the attempt to understand the planets' movements around the Sun. Inspired by William Gilbert's book *De Magnete*, he thought he had found part of the answer in magnetism.

Newton knew about Kepler's three laws of planetary motion: He had studied them at Cambridge. All three of them played a role in Newton's mathematical formulation of the gravitational force. His initial attempt at an explanation depended in part on Kepler's third law, the Law of Harmonics. It states that a relationship exists between the average distance of a planet from the sun and the planet's period—the time it takes for the planet to make one orbit. In mathematical terms, the square of a planet's orbital period is directly proportional to the cube of the planet's average distance from the Sun. So, for example, if Planet B has an orbit four times more distant from the Sun than Planet A, then Planet B's period is eight times larger than that of Planet A. At the same time, its orbital velocity is two times smaller than that of Planet A.

Newton was also familiar with Christiaan Huygens's explanation of "centrifugal force"[1] in *Horologium Oscillatorium*. Huygens discovered that the centrifugal force is directly proportional to the square of a body's velocity and inversely proportional to the radius of the orbit. That means the centrifugal force operating on Planet B must therefore be sixteen times smaller than that operating on Planet A. And if that was the case, then the inward-pulling centripetal or gravitational force must also vary in the same fashion. Newton had figures for the distance of the Moon from the earth and so he tried to deduce the acceleration of the Moon as it "fell" eternally along its orbit around the earth. He came up with a value of 26.3 feet per second each second at the earth's surface. And that was a serious disappointment. For Galileo and others, he knew, had already shown the acceleration of freely falling bodies to be nearly 30 feet per second each second. The difference was too great to be acceptable. Newton set the problem aside for a time and turned to his work on optics and mathematics.

Eventually, though, he returned to the problem of gravitation. By 1686 Newton had solved all the problems he had encountered over the years. The mental effort required, however, wreaked havoc on a mind and body already considerably fragile. He often forgot to eat or sleep, his physical health suffered, and he was prone to long periods of absent-mindedness.

Newton laid out his discoveries in his book *Philosophiae Naturalis Principia Mathematica* (*The Mathematical Principles of Natural Philosophy*). The work he had begun in 1665 on the family farm in

Lincolnshire did not see print until 1687. Perhaps it was a form of stage fright, a fear of being found wrong, or a consequence of his suspicions that others wanted to steal his ideas and work, but Newton was terrified by the prospect of publishing his findings. It took the persistent efforts of his friend Edmond Halley, who financed the publication out of his own pocket, to finally get the *Principia* published.

In the *Principia* Newton revealed how Kepler's three laws led him to formulate the law of gravitation. Newton began by presenting three "Axioms, or Laws of Motion":

- Every body remains in a state of rest or of uniform motion unless it is acted upon by outside forces.
- The rate of change in a body's motion is inversely proportional to its mass and directly proportional to the amount of force applied.
- For every action there is an equal and opposite reaction.

Newton also defined the centripetal force he had previously tried to explain, and offered concrete examples. The one that still captures our imagination today is that of the cannon fired horizontally from a mountaintop. Newton argued persuasively that if the cannonball were imparted a sufficiently high velocity, its curving fall would never quite hit the ground. It would fall *around the earth.* In the same way, he noted, the Moon "falls" around the earth. Newton then used precise and powerful mathematics to show how Kepler's three laws of planetary motion led inexorably to an understanding of the centripetal force that binds the planets to the Sun. Kepler's laws are true because of the existence of a universal attractive force between solid bodies. That force, said Newton, must be inversely proportional to the square of the distance between the two bodies. Double the distance, in other words, and the force of the attraction between the bodies drops to one-fourth of its original strength. Triple the distance and the strength of the force drops to one-ninth.

The moons of Jupiter, discovered by Galileo, are bound to Jupiter by "that very same force which we commonly call gravity," Newton noted. The moons of Saturn, discovered by Huygens, are bound to the ringed planet by gravity. The Moon is bound to Earth by gravity. Earth is bound to the Sun by gravity. And we are bound to Earth by gravity, as is the cannonball fired from the mountain. It is a universal force.

Huygens, however, had a problem with Newton's gravity. And it was a devilish one. He agreed that Newton's gravitational theory was true. But Huygens could not understand *how* gravity connected two bodies. What carried the force from one object to the other? Newton, Huygens wrote, "will have celestial space to contain only very rare matter, in order that the planets and comets meet with less impediment in their course. This rarity accepted, it seems impossible to explain the action of either gravity or light."[2] Huygens saw light as a wave phenomenon, and the medium through which light waves moved across the cosmos was the ether. The gravitational force, too, must have some medium through which to travel. That must also be the ether. But Newton could not allow the ether to exist; if it did, it would impede the movements of the planets around the Sun, and the Moon around the earth, and the Galilean satellites around Jupiter. For Huygens, though, the ether was the absolute *requirement* for gravity to propagate, along with light waves.

The clash of worldviews in this controversy was deep and intractable. Newton himself could not explain the how of gravity. He could not discover gravity's causes and declared, "I frame no hypotheses." Hypotheses, he asserted, had no place in "experimental philosophy." What counted were the mathematics and the observations.

The conflict between Huygens's view of light as wave and Newton's view of light as particle would rage for another 250 years.

Seeing the Light

Meanwhile, astronomers and other "experimental philosophers" moved ahead with the job of uncovering new and exciting views of a cosmos that had hitherto been unknown. One of the tools that opened up new vistas was based on Newton's use of a prism: the spectroscope. The ancestors of the gargantuan high-resolution spectrometer used with the Keck telescope go back to the eighteenth century. The principles of these early instruments would later revolutionize astronomy and give rise to modern cosmology.

The first published use of the word *spectroscope* didn't appear until 1861 (in the *Proceedings of the American Philosophical Society*), and *spectrometer* did not appear in print until 1874. Nevertheless, scientists were using these instruments long before that. In 1752, for example, the Scottish physicist Thomas Melville had discovered that the spectrum of the light created from burning sodium was composed of thin bright yellow bands or lines. By the beginning of the nineteenth century, astronomers and physicists had developed spectroscopes of considerably greater complexity than that used by Melville.

The nineteenth century saw the spectroscope move from curious device to powerful scientific instrument. In 1800 Sir William Herschel discovered that the sun's spectrum extended beyond red into the region of the electromagnetic spectrum today called infrared. A year later J. W. Ritter found that the spectrum also extended beyond violet into the ultraviolet region. The next really significant step occurred when an English doctor, William Hyde Wollaston, built a spectroscope with a difference. Born in East Dereham in England, Wollaston was a doctor with a medical degree from Cambridge. Like many other educated men of his time, Wollaston's curiosity ranged far and wide. He dabbled in many areas of science, including chemistry, physics, and astronomy. He made important contributions to our understanding of electricity, for example, and discovered the elements palladium and rhodium. In 1812 Wollaston invented the camera lucida, a device that uses mirrors to project an image of an object onto a plane surface so its outline can be traced. In doing so he played a small but important role in the development of photography and modern astrophotography.

Wollaston also made a vital discovery about light. Up to about 1800, researchers used a tiny hole in their spectroscopes to channel light to a prism. Wollaston instead used a very narrow vertical slit. This had the effect, when combined with a set of focusing lenses, of creating a finely collimated spectrum with considerable resolution. In 1802 he used his new spectroscope to examine the Sun's spectrum. He saw something no one had seen before: a series of seven dark lines crossing the otherwise continuous band of colored light. Wollaston noted this phenomenon, but he assumed that the bands simply marked the boundaries between the different colors of the spectrum. Wollaston did nothing more to follow up on his observation.

Twelve years later another astronomer did. Joseph von Fraunhofer, a German astronomer, began using diffraction gratings to break the Sun's light into its constituent colors. As noted earlier, a diffraction grating is a glass or polished metal surface with many very fine parallel grooves or slits cut in the surface. It produces optical spectra when light is bent, or diffracted, off the grooves. Fraunhofer used his spectroscope to make a detailed map of the Sun's spectrum. He, too, noted the presence of many black lines across the solar spectrum. Intrigued, Fraunhofer continued his observations. A map he made a year later showed 324 of these lines. What particularly fascinated him was the fact that the lines were always in the same place in the spectrum, and with the same intensities. Fraunhofer had no clue what the lines were, or what caused them, but he knew they meant *something* important. He labeled the most striking ones with letters—A lines, B lines, C lines, and so on. He eventually charted 576 of these lines. He also discovered the same lines, in the same places, in light reflected from planets. In 1823 Fraunhofer observed dark lines in the spectra of stars. Many of them lay in positions different from those in the Sun's spectrum. Fraunhofer concluded that whatever they were, the lines in solar and stellar spectra did not originate in the earth's atmosphere. Their origin must be either in the Sun and stars themselves, or in something lying between them and us.

A brilliant man, Fraunhofer might well have discovered the nature and meaning of the lines in light if he hadn't died in 1826 at the relatively young age of thirty-nine. He never knew what caused his lines, which today are called Fraunhofer lines in his honor. The first clues to their nature began appearing in the years following his death. Inspired by Fraunhofer's experiments and solar spectrum maps, chemists and physicists began using their own spectroscopes to examine the light from just about anything that glowed. They pointed them at flames, sparks, and burning lumps of coal. It soon became clear that two types of spectra existed. Hot, solid objects emitted the first kind, called a continuous spectrum, a continuous band of color, much like that of the Sun. Low-temperature solids produced a spectrum brightest at the red end. High-temperature objects created a spectrum brightest at the blue end. The continuous spectrum had no additional structure like the dark lines in the solar spectrum. Even when researchers added additional prisms to their spectroscopes, they saw nothing more than the continuous band of color.

Light sources like sparks and flames created a second kind of spectrum quite different from the first—a complex mix of light and dark lines where sometimes certain colors might be completely absent. Other sources produced spectra with intensely brightly colored lines. The more prisms researchers added to their spectroscopes, the more detail they could see in these spectra. Researchers found these "line spectra" very confusing. Continuous spectra seemed independent of their source's chemical composition. The continuous spectrum produced by a hot platinum wire was identical to that of a burning chunk of coal. "Line spectra," by contrast, differed from source to source. The spectrum of a spark jumping between two platinum wires differed completely from that of a spark jumping between two rods of carbon. Some sources produced spectra containing brightly colored lines at several different wavelengths. Researchers drawing maps of these bright lines discovered that some seemed identical in position to lines in the solar spectrum. Others, though, were completely different. Other sources produced spectra at only a very few wavelengths and colors. A flame, for example, colored by adding table salt produced a spectrum with only orange-yellow light. This light in turn consisted of just two wavelengths, 589 and 590.6 nanometers (a nanometer is one-billionth of a meter, or 1×10^{-9} meters).

The issue seemed very muddled. No overarching explanation existed to provide a framework for these discoveries. And researchers continued to uncover new, puzzling facts about light spectra. In 1833 David Brewster found that a tube filled with nitric oxide gas blocked several wavelengths of the light emitted by a hot solid object. The following year the British physicist Fox Talbot distinguished the difference between the red spectral lines produced by the flames of burning strontium and lithium. W. A. Miller, in 1845, carried out observations of the spectra of several different metals. The next step in creating a framework came in 1849.

Jean Bernard Leon Foucault is certainly one of the world's most important forgotten scientists. Those who do recognize his name usually associate it only with the Foucault pendulum, which Foucault invented in 1852 and used to definitively prove the existence of Earth's rotation on its axis. He did much more than this, however. He invented the gyroscope (also in 1852), measured the velocity of light with considerable accuracy, and showed that the speed of light in water and other transparent materials decreased in proportion to the material's index of refraction. Foucault's contribution to reading the message of light was just as important. He observed that the D line in the spectrum of an electric arc from burning sodium absorbed the D line in the sun's spectrum.

By 1850 Newton's corpuscular theory of light had fallen into disfavor. Most physicists had become convinced that light, as Christiaan Huygens had propounded, was a wave phenomenon. Light waves had lengths of only a few nanometers or billionths of a meter. They moved at a very high velocity, measured by several researchers at somewhere in the range of 300,000 kilometers per second. Scientists had also established a clear connection between color and wavelength. The longer the wavelength, the redder the color. For example, violet light had a wavelength of about 400 nanometers, green of around 500 nanometers, and red of about 700 nanometers. By the 1830s physicists and chemists had built spectrometers that could distinguish wavelengths of light less than a nanometer in difference. They were no longer measuring differences in visible color, but differences in wavelengths so tiny that the unaided human eye could not possibly tell the difference.

An astonishing change had occurred in humanity's visual perception of reality. Light is the foundation of vision. Without it, we cannot see. Before 1665, not a single human knew that light was anything more than—light. Now, though, some people had begun to see light in a whole new way. It was not something elementally simple. Pure white light was not "pure." It was made of different colors. Nor were colors simple. The color that the human eye perceived as "yellow-orange" could actually be light of several different wavelengths. With a good spectroscope, anyone with a little training could *see* light of 589 nanometers and also light of 590.6 nanometers.

Not everyone would agree with or accept the new explanation of light. In the late eighteenth and early nineteenth centuries, the German scientist and writer Johann Wolfgang von Goethe challenged Newton's vision of light. Goethe believed that white light resulted from an *absence* of all colors and did not, as Newton said, contain all colors within it. Goethe's position was not based on any scientific experimentation or observation, as was Newton's theory. Rather, Goethe based his position on aesthetic and intuitive grounds. Goethe was at heart a Romantic, like his English contemporaries Wordsworth and Coleridge. And like them, he trusted his inner vision and emotional responses to nature. Also like the English Romantic poets, Goethe was something of a mystic. As a young man he had studied not only geology and botany, but also astrology and alchemy. His belief in the mystic properties of light and colors surely influenced his stubborn but futile fight against Newton's vision of light.

And futile it was. Copernicus, Galileo, and Newton had created an alternate astronomical reality for Western culture. Now scientists like Wollaston, Fraunhofer, and Foucault had provided evidence for a new realm of light and had developed instruments that made it possible physically to see this new reality.

The Light from Distant Suns

The work of German scientists Gustav Kirchhoff and Robert Bunsen provided the clue to bring order to the confused line spectrum picture (Bunsen invented a newer type of gas burner, which is still known as a Bunsen burner). The two carried out a series of experiments on flames containing sodium salts. They knew, as Melville had found nearly a hundred years earlier, that such flames emitted a spectrum made of two bright yellow lines. They also knew what Foucault had found a few years earlier: that these two yellow lines seemed to be in the same position as the dark D lines in the solar spectrum. Their experiments eventually showed that any light source, depending on its temperature, could both emit or absorb radiation, creating either bright or dark line spectra. Based on their experiments, Kirchhoff in 1859 announced the formulation of three laws to explain the lines in the spectrum. Today they are known as Kirchhoff's laws.

Kirchhoff's first law basically states that hot, glowing solids, liquids, or gases under high pressure will always produce a continuous spectrum. The chemical composition of the material is not important. Kirchhoff's second law explained the bright line spectrum, today known as an emission spectrum. A luminous gas or vapor under low pressure does not produce a continuous spectrum, but rather a spectrum composed of a series of bright lines. Each element produces its own unique set of lines, its own "fingerprint" pressed into light. Some emission spectra fingerprints are fairly simple ones—the double yellow "D" lines of sodium, for example, seen by Melville in 1752 in the burning sodium flame and by Fraunhofer in 1814 in the solar spectrum. Others are very complex. Iron's emission spectrum, for example, includes hundreds of lines.

The third law presented by Kirchhoff explains both the dark lines in the sun's spectrum and those seen in many laboratories. When a moderately hot gas or vapor lies in front of the very hot source of a continuous spectrum, that spectrum will contain dark absorption lines created by the gas or vapor in the foreground. An easy way to explain this is to describe an experiment carried out by every high school science or physics class. First, we turn on the bunsen burner, introduce some salt into the flame, then examine it with a spectroscope. Voilà: a sodium flame that produces an emission spectrum with two bright yellow lines. Next, we take an ordinary light bulb, turn it on, and examine its light: a standard continuous spectrum. Now, we place the light bulb behind the sodium flame, and examine the flame's spectrum against the light bulb's background light. What we do not see is a continuous spectrum crossed by two bright lines. Instead, we see the bulb's continuous spectrum crossed by two *dark* lines. The reason, said Kirchhoff, is that slightly cooler sodium vapor absorbs the light from the corresponding part of the continuous spectrum. Turn off the light bulb and the bright double sodium lines shine out once more in the same position as the dark absorption lines. Today we know that the atoms in the sodium vapor absorb the radiation coming from the continuous spectrum, creating the dark lines.

Kirchhoff's discovery created a significant revolution in the way astronomers understood the nature of the Sun, and light itself. The Sun must consist of a hot, opaque core surrounded by a cooler atmosphere. The hot inner core created the continuous spectrum; the cooler outer gaseous layers provided the absorption to create Fraunhofer's lines. More than this, though: Astronomers could use a spectroscope to actually determine the Sun's chemical composition. In less than twenty-five years, Auguste Comte's confident assertion that such a task was "eternally impossible" had been turned inside out. Humanity's perception of the universe was beginning to change.

The groundwork completed, astronomers could now make the connection between spectra and spiral nebulae. In 1863 William Huggins attached a spectroscope to his telescope and pointed it at a nebula in the constellation Draco (the Dragon). What Huggins saw surprised him. He expected to see a myriad of lines in the spectrum of the nebula, which would be typical of the spectrum of a star. Instead, he saw one bright line. Huggins first checked his equipment to make sure it wasn't broken. It wasn't. Huggins realized the single bright emission line was real. The nebula could not possibly be a collection of individual stars. It could only be a vast cloud of gas floating in space.

With this one observation the century-long debate about the nature of some nebulae came to a close. Lord Rosse and most other astronomers had been wrong. Not even the most powerful telescopes in the world would ever resolve some nebulae into star clusters. Some nebulae simply were not stars, but clouds of gas.

By the end of the nineteenth century astronomical opinion had shifted. Successive increases in telescopic light-gathering ability and resolution still failed to reveal stars in the spiral nebulae. Astronomers faced a difficult question: Why could they clearly resolve some nebulae into star clusters, but not see *any* stars in the ones that were spiral nebulae? One obvious answer was, if spiral nebulae did in fact contain stars, then they must be far away—*very* far away.

Slipher and Hubble

Vesto Melvin Slipher was another American astronomer born in the Midwest—in his case, Mulberry, Indiana, in 1875. Slipher spent nearly all his professional life at the Lowell Observatory near Flagstaff, serving as acting director from 1916 to 1926 and as director from 1926 to 1954. There he mastered the scientific discipline of spectroscopy and used the spectrograph to make numerous important discoveries about our neighboring planets and distant stars and galaxies. For example, he determined the rotation rates of Venus, Mars, Jupiter, Saturn, and Uranus. Slipher also showed that the atmospheres of Saturn and Jupiter consist largely of methane and ammonia. Under his direction, in 1930 a young farm boy named Clyde Tombaugh discovered the solar system's ninth planet, Pluto. Slipher did all this with his feet on the ground in the mountains of northern Arizona; it was decades before the advent of the robot space probe and the planetary flyby.

In 1909, using an excellent 60-centimeter refracting telescope at the Lowell Observatory in Arizona, Slipher began a research program on spiral nebulae. Like many other astronomers at the time, Slipher believed spiral nebulae were objects lying within the Galaxy, probably other solar systems in the process of forming. His plan was to use a spectrograph to read the lines of light from the spirals. Slipher figured the information gleaned from the light of the spiral nebulae would actually provide some clues to the origin of our own solar system. In 1914 he aimed his telescope at the Andromeda Nebula, and eventually got four excellent photographic plates using the spectrograph. He soon noticed that the spectral lines in the light from Andromeda were shifted from their standard positions.

Slipher knew of several possible causes for such shifts in spectral lines. The simplest explanation, though, was that the object emitting the light—in this case, the Andromeda Nebula—was moving toward the observer. When an object emitting light is moving toward an observer, the lines in the object's spectrum will shift toward the blue end of the spectrum. If the object is moving away from the observer, the lines will shift toward the red end of the spectrum.

Such displacements of the spectral lines are called Doppler shifts, the apparent change in wavelength of radiation emitted by a moving body. A classic example of a Doppler shift is that of a car whose driver is tooting her horn as she drives past you. As the car approaches, you will hear the horn's sound rise in pitch. As the car passes you the sound of the horn will seem to drop in pitch.

Sound is a wave phenomenon. It is caused by pressure waves passing through the air, waves made by the compression and expansion of air molecules. The sound's wavelengths shorten as the car approaches you, and so the pitch rises. The wavelengths of sound lengthen out as the car moves away from you, and so the pitch drops. The nineteenth-century Austrian physicist Christian Johann Doppler first described and explained this phenomenon, thus earning himself a measure of fame and innumerable short biographical listings in reference books.

Light, as we've seen, also often acts as if it were a wave. And because they are analogous to a sound's shift in pitch, the shifts in spectral lines of the kind discovered by Slipher in the light from the Andromeda Nebula are also called Doppler shifts. Slipher could easily calculate the velocity of Andromeda from the amount of blueshift in its spectral lines. The figure he arrived at was surprising, to say the least. Andromeda appeared to be approaching Earth at a velocity of 300 kilometers per second. By the summer of 1914 Slipher had measured the velocities of a total of fifteen different spiral nebulae. Unlike Andromeda, most of them had redshifts in their spectra rather than blueshifts. Slipher assumed that the redshifts he was detecting were Doppler shifts and that the objects were receding from Earth. But other possible causes existed for the redshift in the light from these spiral nebulae. For example, clouds of gas and dust lying between the earth and the nebulae would have the effect of reddening any light passing through them. But no evidence existed for any such dust and gas clouds between all the nebulae and Earth. The simplest explanation was that they were moving away.

And at fairly astonishing velocities, too. The fastest one had a redshift that translated into a recessional velocity of 1,100 kilometers per second. Nothing within the Milky Way Galaxy had ever been found to be moving at such a speed. In fact, anything moving that fast would escape the gravitational pull of the entire Galaxy. Slipher's unexpected discovery was one of the most surprising findings of recent times.

Slipher announced his news in August of 1914 at a meeting of the American Astronomical Society, triggering another rumble through the supposedly solid ground of the Western world's cosmological landscape. One of the people in attendance was the young Edwin Hubble. He was already aware of the arguments that spiral nebulae were not located within the Milky Way, but were "island universes" lying far beyond it. Slipher's announcement may well have spurred his interest in nebulae. In any case, he was certainly interested in the nature and location of nebulae even at this point in his life. Hubble's doctoral dissertation, completed while he worked at Yerkes Observatory as a Ph.D. student, was entitled *Photographic Investigations of Faint Nebulae.*

World War I interrupted Hubble's journey into astronomy and cosmology. After completing his dissertation he joined the American Expeditionary Force as an ambulance driver in Europe. When he returned home, however, he picked up where he left off. In 1919 he journeyed west from Chicago and the Yerkes Observatory site in Wisconsin. He headed toward a spot high in the mountains outside of Los Angeles. The Mount Wilson Observatory was the site of the 1.5-meter Ritchey reflecting telescope and the then-giant 2.5-meter Hooker reflector.

Mount Wilson itself exists because of earthquakes. A piece of the earth's crust called the Pacific Plate, along with much of California, is slowly sliding northwest along a boundary with the North American Plate, which underlies most of North America. The troubled edge between the two is the San Andreas Fault. As the plates grind past one another, earthquakes along the San Andreas and other faults have pushed up the San Gabriel Mountain Range. The 1,741-meter-high (5,710 feet) Mount Wilson is part of that range, lying a few kilometers north of the well-to-do community of Altadena.

If California's earthquakes bothered Hubble, it wasn't enough to send him back to the Midwest. Mount Wilson was in those days the Mecca of contemporary astronomy. And Hubble was going where the action was.

For the first several years of his stint on the Mount Wilson staff, Hubble spent a lot of time examining the giant galaxy known as M87. Today astronomers know that this galaxy is what is called an elliptical galaxy, because of its shape and the kinds of stars it contains. In the early 1920s, though, many astronomers thought M87 was one of the spiral nebulae. Hubble was already fascinated by the cosmological puzzle of the spirals, and M87 was one on which he concentrated his attention. Using the superb 2.5-meter Hooker and the latest available photographic plates, Hubble was able to get pictures of starlike condensations surrounding M87. Were they indeed stars? And if so, what did that say about the nature of spiral nebulae?

If they were stars, Hubble could compare their apparent brightness with the brightness of stars known to be within our own Galaxy and thus make a good guess at the distance of M87. It appeared that this "spiral nebula" lay outside our Galaxy. But Hubble was not positive that the starlike images he photographed around M87 really *were* stars. He needed some solid proof. He needed the information carried within the light and deciphered by instruments such as spectrometers. That kind of data would truly reveal if the objects were really stars. As it happened, Hubble was not examining a spiral nebula, but a globular cluster. As we saw earlier, globular clusters are "local objects"; they are gravitationally bound to our own galaxy in elongated orbits. But the images he photographed, and the difficulty he had in identifying them, pushed Hubble into another round of investigations.

In 1923 Hubble began studying the distribution of galaxies in space and their distances from the earth. He was particularly interested in studying the appearance of novae in spiral nebulae. A nova (from the Latin word meaning "new") is a star that suddenly increases in brightness for a time, and then dims again. For a few days or weeks a nova becomes as bright as all the stars in our Galaxy, and then fades to invisibility. About ten to fifteen novae occur each year in the Galaxy. Most are believed to be close binary stars, two stars orbiting one another. One of the pair is a tiny white dwarf star while the other is a bloated red giant. Matter blown off the red giant star falls onto the surface of the white dwarf, where it builds up and finally explodes like a superhydrogen bomb. The result: a flare-up of light and energy we call a nova.

Some "novae" are hundreds to thousands of times more energetic than most others. Today we know that these particular novae, called supernovae, are old bloated stars that have collapsed and then exploded.

Still other novae brighten and dim on very regular schedules, and are more properly called variable stars. The periodic brightening and dimming of variable stars has several different causes. Some, for example, are actually binary stars where one star is dimmer than the other. If their orbits lie in our line of sight, the two stars will periodically eclipse one another. When one star moves in front of the other, the light we see from both seems to dim. This dimming is particularly pronounced when the fainter star moves in front of the brighter companion. The classic example of such "eclipsing variables" is the star Algol, the "Demon Star" of medieval Arab astronomers. Actually two stars closely orbiting one another, Algol appears to the naked eye as one star. Its regular "winking" in brightness is the result of this particular eclipsing pattern.

Other variable stars are not binary, however. They are individual stars that for any of several reasons brighten and dim with regularity. One particularly important kind of variable star played a key role in Hubble's cosmological breakthrough, and thus in the reconfiguring of our vision of reality.

Two other important contributors to twentieth-century astronomy, and to the revolution in our vision of the cosmological landscape, were Edward Pickering and his younger brother William. Edward, for example, in 1882 had developed a way to photograph the spectra of several stars at once by placing a large prism in front of a photographic plate that was exposed to the light from the target stars. He also discovered the existence of spectroscopic binary stars—stars orbiting one another so closely that only the alternating shifting of their spectral lines revealed their dual nature. Brother William in 1899 discovered Phoebe, one of the moons of Saturn.

In 1912, while Hubble was still a Rhodes scholar in England, Edward Pickering was the director of the Harvard College Observatory. Along with William he led a team of astronomers and technicians observing stars in the two Magellanic Clouds. The Magellanic Clouds are actually two of several small satellite galaxies of our own Milky Way. However, in 1912 no one was really sure of that.

In fact, no one really knew if anything existed beyond the Milky Way. Most astronomers believed that every object seen in the night sky lay within the Milky Way Galaxy. The Milky Way *was* the universe, as far as nearly everyone was concerned. The universe was therefore a fairly small entity. True, it was far larger than the crystalline spheres of the ancient Greeks; larger than Ptolemy's universe of equants and epicycles; larger even than the cosmos revealed by Copernicus and Galileo. But it was still a *tidy* universe, one where your neighbors weren't all that far away. It was, in other words, a universe much like the social and cultural landscape in which most Europeans and Americans lived at the end of the nineteenth century.

When it finally changed, it would become a universe much bigger, emptier, and lonelier than anyone could conceive.

One of America's most important astronomers of the early twentieth century, Henrietta Swan Leavitt, worked on the Pickerings's Magellanic Cloud team. One of her tasks was observing a class of variable stars called Cepheid variables in the Small Magellanic Cloud. Cepheids are so named in honor of the star Delta Cephei in the constellation Cepheus, the first variable star of its kind to be identified. In 1895 Solon I. Bailey had discovered Cepheid variables in the outer regions of several globular clusters. Not long afterward Cepheids were also found in the two Magellanic Clouds. The brightness of Cepheids varies because the stars pulsate, increasing and decreasing in radius by as much as 10 percent. Cepheids vary in brightness on a highly regular basis.

In 1912 Leavitt made an important discovery about Cepheids in the Magellanic Clouds. She found that a Cepheid's greatest brightness was directly related to its period of variability. The longer its period of pulsation, the brighter it became. This in turn meant that Cepheids could be used as "standard candles." If one could measure a Cepheid's pulsation period, then one could also deduce the star's absolute magnitude or brightness. By comparing the absolute magnitude with its apparent magnitude in the night sky, one could then determine the distance to the star. That would then make it possible to use Cepheid variables as cosmic yardsticks.

All of this was predicated on knowing the actual distance from Earth of at least one Cepheid variable—establishing at least one "zero point." The principle was simple, since a standard formula exists for the relationship of a star's absolute magnitude, apparent magnitude, and distance from Earth. Suppose, for example, that a Cepheid variable star with an apparent magnitude of 2 (fairly bright) is found by independent means to be 61 light-years distant from Earth. The star's absolute magnitude is therefore −4.5. Knowing this and the relationship between a Cepheid's pulsation period and its apparent brightness, an astronomer can quickly determine the distance of *any* Cepheid. (In fact, the example above is an actual Cepheid variable. Polaris, the North Star, varies in brightness by one-tenth of a magnitude every four days.)

In 1913 astronomer Ejnar Hertzsprung was able to establish the zero points for thirteen Cepheids within the Milky Way. He found that an absolute magnitude of −2.3 for a Cepheid corresponded to a period of 6.6 days. At the same time a young and ambitious astronomer named Harlow Shapley also determined the distance to about a dozen Cepheids located in several globular clusters. Leavitt took these findings and applied them to the Cepheids in the Small Magellanic Cloud. Her calculations revealed that the Magellanic Clouds were at least 35,800 light-years distant. This still was within the confines of the Milky Way Galaxy as it was understood in 1913. They would not remain there for long, however.

Like Hubble and Slipher, Harlow Shapley was a product of the Midwest, having been born in Nashville, Missouri. From 1914 to 1921 he was on the staff at Mount Wilson. He then became director of the Harvard Observatory, where he worked until 1952. Shapley had first determined the distance to Cepheids in about a dozen different globular clusters. In 1916 and 1917 he made a series of photographs of globular clusters using the 1.5-meter Ritchey reflector at Mount Wilson. Shapley began to see that the brightest Cepheids in each cluster all had about the same absolute magnitude. It also became clear that the globular clusters were all pretty much the same size. Their overall brightness in the sky was a direct result of their distance from Earth. Shapley could now extend his distance measurements. He proceeded to use Cepheids to measure the distances of the sixty-nine globular clusters then known.

To the surprise of many, Shapley found that globular clusters themselves were arranged in a sphere about 260,000 light-years in diameter. The center of the sphere was in the constellation Sagittarius. This raised another question: Why were globular clusters not distributed equally in the night sky but instead squished into only half? Shapley at this point took a great intuitive leap. Suppose (he said) that the sphere of globular clusters was at the center of the Galaxy? If it were, then the visible distribution of globular clusters could only mean that (1) the solar system did not lie near the galactic center and (2) the Milky Way Galaxy itself must be much larger than anyone had hitherto suspected.

Shapley also took another look at Leavitt's figures for the distance to the Magellanic Clouds. He discovered an error in the brightness scales for some of the stars in the Southern Hemisphere. Shapley redid the calculations for the distance to the Cepheids in the Magellanic Clouds. The result was astonishing: the Magellanic Clouds were at least 94,000 light-years distant, well beyond the boundaries of the Milky Way. Today we know that the Large Magellanic Cloud is about 163,000 light-years distant, in the direction of the constellation Dorado (the Swordfish) in the sky of the Southern Hemisphere. It has a mass of about ten billion suns and is about 39,000 light-years in diameter. The Small Magellanic Cloud is about 20,000 light-years in diameter, with a mass of about two billion suns. It is nearly 196,000 light-years distant, in the direction of the Southern Hemisphere constellation Toucan.

Bolstered by his findings—and with the brashness for which he was famous—Shapley boldly announced his estimate for the size of the Galaxy: 300,000 light-years in diameter. Astronomers were appalled by this figure and rejected it as far too large for several reasons. For one thing, Shapley had fashioned his model of the Galaxy using a new and relatively untried technique of his own devising. Astronomers were suspicious of its accuracy. Also, considerable observational and statistical evidence already existed that the Galaxy was much smaller than 300,000 light-years.

The first modern estimate of the Galaxy's size and shape had come from the single most famous astronomer of the late eighteenth century. Sir William Herschel had discovered the planet Uranus in 1781. His real love, though, was stars, particularly double stars. As we've seen, a double, or binary, star is simply two stars in orbit around one another. Herschel had discovered their existence in 1804, applying Newton's law of gravitation to explain their motions in the night sky. Herschel spent many years finding and cataloging double stars. As part of that project he decided he needed to know the overall distribution of stars in the universe. In 1784 he developed a method he called "star-gauging." Herschel would point his telescope at specific areas of the sky, and then carefully count (or "gauge," as he called it) all the stars he could see in the telescope's field of view. He surveyed and counted the stars in more than fourteen hundred different areas of the sky. Over the following years Herschel published some of his results. The picture that emerged from his star-gauging was of a stellar system in the shape of a very rough disk. It became known as the "grindstone model" of the universe. He calculated that it was five times as wide as it was thick, and that one side was split into two branches divided by a space almost completely free of stars. The Sun and its planets, Herschel concluded from his stellar survey, lay near the center of the disk. There was no way to say exactly how large the disk actually was, since no one then knew how to determine the distance to other stars. But it was a start. The Galaxy, it seemed, looked like a big fat lens.

This stellar grindstone was believed to be the entire universe. In the late eighteenth century no one knew about things like external galaxies, redshifts, the expanding universe, galactic superclusters, or the Big Bang. What could be seen constituted the universe, and it all lay within the Milky Way Galaxy.

By the first decade of the twentieth century, the astronomical community generally accepted the idea that the Galaxy had some kind of flattened structure. It probably looked a bit like Herschel's grindstone, or perhaps like a flattened sphere. Detailed star counts and careful mapping of stellar distribution seemed to point to this kind of structure.

More than a hundred years after Herschel announced his grindstone model of the universe, the Dutch astronomer Jacobus Cornelius Kapteyn started carrying out his own version of star-gauging at his observatory in Groningen. Since Herschel's time two important discoveries about the Galaxy had been made. In 1904 J. Hartmann had discovered the presence of interstellar gas in the regions between the stars. Five years later E. E. Barnard discovered the existence of interstellar dust. Astronomers could now distinguish between those regions of the sky that really had few stars and those dark regions where clouds of dust and gas obscured the view of stars beyond. Kapteyn began surveying the sky, using photographic techniques to get detailed star counts in selected areas. Over the years a new picture of the Galaxy began to emerge. In 1922 he finally published the results of his work, a model of the Galaxy now known as the "Kapteyn universe." This model depicted the Galaxy as shaped like a cushion, packed with stars at the center and thinning out toward the edges. Kapteyn's model was about 5,200 light-years in diameter and about 1,000 light-years thick. It contained about 80 billion times the mass of the Sun, and about 50 billion stars. The Sun was about halfway from the top and bottom and some 200 light-years from the center.

The Kapteyn universe was very much smaller than Shapley's model of a galaxy 300,000 light-years across. It was also smaller than the model commonly held by most of the astronomical community. Its advantage was that it clearly implied that both globular clusters and spiral nebulae lay outside the Galaxy.

The Great Debate

By 1919, then, the astronomical community had a bit of a controversy on its hands. The work by Kapteyn and others seemed to suggest strongly that the Milky Way Galaxy was 5,200 light-years across. Spiral nebulae like Andromeda therefore lay far outside the confines of the Milky Way. They must be "island universes." Harlow Shapley's Cepheid counts, however, suggested that the Milky Way was 300,000 light-years in diameter. Spiral nebulae, therefore, were nothing more than nearby nebulous objects, just as Lord Rosse and others had once supposed. Finally, many astronomers held to a third, intermediate viewpoint. In this cosmological picture, the Milky Way Galaxy was spiral in shape, somewhat larger than the Kapteyn model, but considerably smaller than Shapley's suggestion. Spiral nebulae in this model did indeed lie beyond the borders of the Milky Way—but not as far as some would suggest.

At the end of 1919, as Edwin Hubble was beginning his long career at Mount Wilson and starting to focus on M87, George Ellery Hale made a fascinating suggestion to the National Academy of Sciences. Perhaps the Academy ought to devote one of the upcoming lectures in memory of Hale's father to the island universe controversy or, perhaps, to relativity. The Academy decided to go with island universes and picked two speakers to present the major views in the controversy—Harlow Shapley and Heber Curtis. An astronomer at Lick Observatory, Curtis was a tenacious defender of the island universe theory.

The "great debate" between Shapley and Curtis took place on April 26, 1920, at a public meeting in Washington, DC. However, the event wasn't a debate at all. In fact, the two principals gave two entirely different kinds of presentations. Shapley spoke first, and gave a popular-style talk that defended his theory that the Milky Way was a giant galaxy. Curtis had come prepared to give a formal presentation, with charts, graphs, and slides, defending his island universe theory and attacking Shapley's proposal. He was quite surprised, then, when Shapley launched into an exposition that was quite elementary in tone. Still, when his turn came to speak, Curtis pressed forward and gave his talk as he had planned it. There was no debate, and questions from the audience were taken.

Despite the unexpected dissonance between the two presentations, everyone at the lecture agreed that it had come off swimmingly. Both Curtis and Shapley had made their points with clarity and force. In fact, organizers for the 1921 meeting of the Astronomical Society of the Pacific asked the two to repeat the performance for that gathering. Shapley and Curtis declined.

The irony of the great debate of 1920 is that both Curtis and Shapley were right—and wrong. Curtis's "island universe" turned out to be correct. But so was Shapley's contention that the Milky Way Galaxy was much larger than most astronomers then believed, and larger even than Shapley himself assumed.

However, Shapley was wrong in his assertion that spiral nebulae were contained within our Galaxy. Three years after the event, Edwin Hubble would write Shapley a letter containing the information that would prove him wrong.

Hubble's Vision

Our current vision of the architecture of the universe springs from the photographs and spectrographs created by Edwin Hubble. His hope, as he began a survey of Andromeda in 1923, was to find some novae in that spiral nebula. If he could find and record enough of them, he felt, he could make some good estimates of their apparent brightness. Then he would compare the apparent brightness of novae in Andromeda with the known absolute brightness of novae in our own Galaxy. Hubble might then be able to estimate the distance to the Andromeda Nebula with a fair amount of certainty.

In October 1923, however, Hubble made a major discovery. He spotted a Cepheid variable star in Andromeda. Early in 1924 he found two more. Hubble worked out the periods for the Cepheids and then calculated their estimated actual magnitudes. From those he worked out the estimated distance from Earth to the Andromeda Nebula. It was nearly a million light-years.

Hubble wrote a note to Shapley early in 1924, telling him of his discovery and enclosing a hand-drawn light curve on graph paper for the first Cepheid he had found. It was the death knell for Shapley's proposal that spiral nebulae lay within the Galaxy. Even if the Milky Way were 300,000 light-years in diameter, Andromeda still lay far beyond it.

Between 1923 and 1925 Hubble took telescopic photographs of 1,283 sample regions of the sky and counted the numbers of galaxies in each region. From these observations he concluded that on a very large scale the distribution of galaxies in the universe is uniform. What Hubble had detected was observational evidence for the "cosmological principle." The cosmological principle is an essentially philosophical statement that at any point in time the universe on the large scale looks the same no matter who observes it from any vantage point. In other words, the universe is pretty much the same anywhere.

The cosmological principle's one big problem is that it doesn't appear to be entirely true, at least not from our vantage point on Earth. On the scale of the entire universe, however, it appeared to Hubble that the cosmological principle was true. Large-scale isotropy (properties having the same values no matter where they are measured) and homogeneity apply to the entire universe, said Hubble. But on a smaller scale the universe looked decidedly lumpy. Galaxies have a tendency to cluster together. Astronomers already knew of the existence of a few galaxy clusters in 1924. And contrary to his own intentions, Hubble found a lot more. He had picked his sample regions with an eye to deliberately avoiding those with galaxy clusters. But to no avail. There they were, almost everywhere he looked. Hubble concluded that most if not all galaxies are either members of clusters and groups or were originally formed in one.

Other observers would later find conclusive evidence that galaxies seem to cluster together. Not long after World War II, astronomer George O. Abel compiled a catalog of 2,712 galaxy clusters and groups. Out of that he filtered a homogeneous sample of 1,800. He, too, concluded that on the very largest scale the galaxy clusters themselves had a homogeneous and isotropic distribution. Once again, here was confirmation of the cosmological principle. But on a smaller scale, Abel found more complexity and inhomogeneity.

The relationships of these different macrocosmic structures resemble that of a house to a group of cities. The house is part of a neighborhood, a collection of houses connected by a network of local roads and streets. Various neighborhoods make up a town or city. A small town may really be nothing more than a neighborhood, whereas a large city, or metropolis, contains dozens—perhaps even hundreds—of neighborhoods, plus industrial areas, shopping malls, parks, a government center, and perhaps a university with its associated student and faculty housing. Finally, some cities and towns often cluster together. Some such clusters may be large enough to be called a megalopolis, or "strip city"—the Boston–Washington megalopolis, for example, or the strip city that runs from Seattle to Olympia, Washington.

We can compare this civic structure to that of the universe. First of all, the "house" we call the solar system—the Sun and its retinue of planets (including Earth)—is part of a "neighborhood." The neighborhood we live in is a moderately large spiral-shaped galaxy we call the Milky Way. We are located at the outer edge of the neighborhood, on one of the Galaxy's spiral arms, about 29,000 light-years from the bulging galactic center. Considerable evidence now exists that the galactic center is dominated by a supermassive black hole, with a mass of at least a million suns. The Galaxy itself includes at least 200 billion stars, and probably more, with additional huge amounts of matter in the form of gas and dust clouds. About 90 percent of the Galaxy's mass appears to be invisible, however. We can neither see it with our unaided eyes or optical telescopes nor detect its presence from any emission of electromagnetic radiation. This "dark matter" emits no radio waves to "hear," no infrared radiation to "feel," and no X rays, ultraviolet rays, or cosmic rays to detect with orbiting observatories. Only our observations of the gravitational interactions of galaxies and galactic superclusters reveal the presence of this dark matter.

The Milky Way Galaxy is gravitationally bound to several neighboring galaxies. The two best-known ones are the Small and Large Magellanic Clouds (abbreviated SMC and LMC). Both can be seen only in the night skies of the Southern Hemisphere. The LMC is the site of a recent supernova, 1987A, the first exploding star in our galactic neighborhood to be seen in more than three hundred years. The SMC and LMC are so close to the Milky Way Galaxy, in fact, that some astronomers think they may someday merge with it. Others think the two star clouds were recently torn from the larger Milky Way by some still unknown gravitational interaction with another large galaxy. Seven other galaxies, all quite small, also lie relatively near the Milky Way Galaxy.

The Milky Way and its companion galaxies are in turn part of a "cosmic city," a larger cluster of galaxies. This city of galaxies covers about 3 million light-years of space, and the galaxies themselves are gravitationally bound together. How do we know this? Using spectrographs based on Newton's prism and light beam, astronomers can measure the shifts in the spectral lines first discovered and explained by Wollaston and Fraunhofer. Those shifts tell astronomers how fast the galaxies are moving and in roughly what direction. By combining that with other clues to the total mass of different galaxies, they can determine if the galaxies in some region of space are bound together by the invisible threads of gravity.

Given enough time, the galaxies in this group will eventually collapse together. Astronomers today refer to such "dynamically collapsed systems," which contain a few dozen to several hundred galaxies, as clusters. Structures containing no more than a few dozen or so galaxies are called groups. The Local Group, the "galactic city" that includes the Milky Way, contains at least twenty-eight galaxies, most of which are smaller than the Milky

Way. The other major member of the Local Group is M31, the Andromeda Galaxy, about 2.4 million light-years distant from us. Andromeda is slightly larger than the Milky Way. It, too, has an entourage of smaller satellite galaxies. The companions M32 and NGC 205 are clearly visible in any good photograph of Andromeda. At least seven other small galaxies in the Local Group are bound to M31 by gravitation.

Besides the Milky Way and Andromeda, the Local Group includes two intermediate-sized galaxies. The Large Magellanic Cloud is one. The other is the well known Triangulum Nebula (also known as M33), which lies less than 650,000 light-years from Andromeda, and may be gravitationally bound to it. The other members of the Local Group are all so-called dwarf galaxies.

The next step up in structural scale includes clusters of clusters and groups—the "metropoli" of the universe. Three other galaxy groups lie close to the Local Group: the Sculptor Group, which includes NGC 253; the M81 Group; and the Maffei 1 Group. Along with these and other groups and clusters, the Local Group is a member of the Coma-Sculptor Cloud. This collection of galaxy groups and clusters includes several hundred galaxies. The Coma-Sculptor Cloud stretches out in a pancakelike shape for more than 40 million light-years.

The Coma-Sculptor Cloud, in turn, is part of a still-larger structure called the Local Supercluster, a megalopolus of the universe, as it were.

In his galactic survey, George Abel identified seventeen distinct galactic superclusters. Since then, astronomers have confirmed that galaxy groups and clusters throughout the universe group together in superclusters. The Local Supercluster is about 100 million light-years across. The center of the Local Supercluster appears to be in the direction of the constellation Virgo, about 30 to 50 million light-years distant. Typically, superclusters are about 100 million light-years in diameter. They contain one or two very large galaxy clusters on the average, though some superclusters include 100 or more galaxy clusters.

At this point it appears that galactic superclusters are the largest sets of structures in the universe. The hierarchy ends there. So the universe appears lumpy up to a scale of about 100 million light-years or so. Above that, the cosmological principle seems to hold sway.

In the 1920s, however, the discovery of galactic superclusters still lay years in the future. What cosmologists knew for sure (more or less) was that stars congregated in galaxies, either individually or in star clusters of various kinds; that galaxies tended to cluster together in larger groups; and that the universe was obviously much larger than just the Milky Way Galaxy.

The Expanding Universe

In 1687 Isaac Newton (with the prodding of his good friend Edmond Halley) had finally published his *Principia*. In the *Principia* Newton had propounded his famous law of universal gravitation. In the process of doing so he considered the effects of gravitation on the universe at large. It was a difficult matter for Newton to consider, for it forced a potentially serious conflict with his religious faith. A large percentage of scientists today, including astronomers, consider themselves religiously unaffiliated. But this was not at all the case in the seventeenth century. Christianity was vigorously contending with the rising rationalist philosophies for the minds and hearts of the people. Most natural philosophers (which is what scientists were referred to back then) were devoutly religious men.

Isaac Newton was no exception. He truly believed that God was infinite in all respects and that nothing else could be. There was one teeny problem, though. If the universe was *not* infinite in size, then the force of gravity would cause all the objects in the universe eventually to collapse together. Newton looked out and up and around, and it seemed manifestly clear to him that the cosmos was not collapsing in upon itself. Suppose, then, that the universe *was* infinite. Ah, then, the gravitational potential would be the same everywhere. An infinite universe could never collapse in upon itself. Smaller parts of it could, wherever irregularities in the density of mass concentrations existed. But the entire infinite universe itself, and everything in it, would be safe from such a collapse. Newton refused, however, to concede that anything other than God could be infinite. He more or less concluded that the universe was finite, but didn't collapse because God didn't let it. The universe was thus static in Newton's cosmology, neither growing nor shrinking, with gravity governing the smooth movements of the stars and planets in their courses. Newton's mathematical proposition that the universe was infinite, however, remained in his writings.

Nearly 230 years later, in 1916, Albert Einstein published his general theory of relativity. General relativity is a theory of gravitation that includes Newton's gravitational theory as a valid but special-case subset. One consequence of general relativity is a series of mathematical equations that eliminate any possibility of a static, stable universe. Whether it was finite, as Newton religiously insisted, or infinite, the universe could not be static and stable. It was either expanding or contracting. And what possible force could exist in the cosmos to counteract the relentless pull of gravity?

General relativity's equations treat gravity as a geometric "warp" in space-time. The presence of matter in the cosmos causes space-time near it to "bend." That bending we perceive as the force of gravity. The bigger the piece of matter, the steeper the warp in space-time and thus the stronger the gravity field surrounding it. Matter and energy are also equivalent, according to Einstein's earlier special theory of relativity (yes, the famous equation $E = mc^2$). The universe is filled with matter and energy, and all of space-time is curved. All objects in the universe—from photons of light to the galaxies themselves—move through the cosmos along curved, unaccelerated paths called *geodesics*.

Like Newton, Einstein couldn't stand the idea of a universe that would some day collapse in upon itself. Newton rejected the conclusions of his own equations because of his religious beliefs. Einstein did the same because of his firm belief in the "common-sense" cosmology of his day, the cosmology of a static universe. So Einstein fudged a bit. He inserted into his equations a mathematical term he dubbed the "cosmological constant." A positive constant of just the right value implied the presence in the cosmos of a universal *repulsive force* that exactly balanced the universal attractive force of gravity.

There was nothing mathematically wrong with Einstein's cosmological constant, but there was no observational evidence for its existence and nothing like this universal repulsive force existed in Newton's gravitational theory. Einstein had not introduced it into his equations for any scientific reason, but solely to save his own cosmological vision.

As soon as Einstein published his general theory of relativity, it moved into the larger communities of physics and cosmology. Other scientists beside its originator began looking at it, studying it, and pushing, poking, and prodding at its equations. They began finding implications and conclusions that Einstein himself had not seen. Late in 1916, for example, Karl Schwarzchild developed a series of equations based on Einstein's that revealed the theoretical possibility for the existence of objects that are today called black holes. In the 1920s other physicists and cosmologists looked at general relativity and did not see a problem with a static universe. The field equations of general relativity, in their simplest form without a cosmological constant, could apply equally well to a universe either uniformly contracting or expanding. In 1919 Willem de Sitter proposed a cosmological model of an expanding universe based on Einstein's general relativity equations. Three years later Russian mathematician Alexander Friedmann proposed a theory that improved on de Sitter's model.

In 1927 the Belgian priest and cosmologist Georges Lemaître, unaware of Friedmann's work, came up with a nearly identical model of an expanding universe. Lemaître did Friedmann one better, too: He proposed some astronomical observations that could test the validity of his theory. Lemaître knew about the observations of the spectra of spiral nebulae that American astronomer Vesto Slipher had made in 1912. He knew that Slipher had found that they often had Fraunhofer lines displaced toward the red end of the spectrum and that this implied the nebulae were moving away from Earth. Lemaître thought that further observations of the kind made by Slipher might provide strong observational evidence that the universe really was expanding.

In hindsight it is now clear that Lemaître's 1927 paper on an expanding universe was seminal. It broke new ground in cosmology. At the time, however, almost no one paid much attention to it.

Lemaître later remarked on a conversation he had with Einstein shortly after the paper had been published. Einstein had told him, "Your calculations are correct, but your physical insight is abominable." In fact, it was Einstein who had made the colossal blunder. He himself would later call his invention of the "cosmological constant" the single most important mistake he had ever made. Lemaître's insight into the nature of the cosmos turned out to be at least as correct as his mathematics.

Slipher had faced a significant difficulty in accurately interpreting his observations. In 1912 not everyone believed that spiral nebulae were really separate galaxies, and no one knew how far away they were. In 1927, when Lemaître first published his proposal of an expanding universe, astronomers and cosmologists had been talking for nearly a decade about the implications of galactic redshifts. But the observational evidence for an expanding versus static universe was still weak. Edwin Hubble changed that.

In 1929 he made another discovery that led to a revolution in cosmology. That year he combined new measurements of distances to other galaxies with earlier and more recent observations of their spectra. The galaxies he observed had significant redshifts in their spectra. The more distant the galaxy, the larger was the redshift. What Hubble had done was reproduce on a much larger scale the first, limited findings of Vesto Slipher some fifteen years earlier. Hubble had been a young graduate student then, sitting in the audience as Slipher made his announcement. Now Hubble was playing the role of teacher and leader.

Two possible explanations existed for this redshifted state of affairs. One was quickly proposed by Swiss astronomer Fritz Zwicky. He suggested that the redshifts of distant galaxies were caused by the "aging" of the photons of light coming from them. As they made their long journey through space to finally reach Earth, the photons passed through intervening clouds of dust and gas. Their collisions with dust particles and gas molecules caused the photons to lose energy. This in turn caused their wavelengths to lengthen and their spectra to redden. The more distant the galaxy, the more gas and dust its photons had to plow through to get to Earth, and thus the redder they would appear.

The problem with this explanation was that there was no other evidence for the existence of huge amounts of dust and gas between the galaxies. All astronomical observations appeared to indicate that intergalactic space was pretty much empty. Zwicky's hypothesis made sense, but had no observational foundation. It quickly fell into disfavor.

Hubble offered the other explanation, and it was much simpler. The redshifts, suggested Hubble, were caused by the motions of the galaxies themselves. Galaxies moving toward us would have their light shifted toward the blue end of the spectrum; those moving away from us would have their light shifted toward the red end. No matter where Hubble looked in the sky, almost all the galaxies had redshifts. The only exceptions were several galaxies like Andromeda, lying close to the Milky Way Galaxy and obviously bound to it gravitationally. All the others, though, had redshifts. They were all, said Hubble, moving away. By choosing the Doppler shift explanation, Hubble had taken the tried-and-true philosophical path followed by science as a whole, and laid down in the early fourteenth century by William of Occam. Known as Occam's razor, this principle states that "entities must not be needlessly multiplied." In other words, when several explanations for a phenomenon exist, the simplest is most likely to be correct.

What Hubble (and all other astronomers and cosmologists) did not for a second suggest was that the earth and its galaxy were actually lying at the center of all this galactic recession. Copernicus had long ago driven the spike through the heart of cosmological geocentrism. Moreover, the cosmological principle itself precluded such a privileged position for Earth. Everywhere in the universe is pretty much like everywhere else, no matter what the time or location of the observer. What we see, then, is what an observer in the Andromeda Galaxy, or in M81, or in the Triangulum Nebula, or anywhere else would see. They, like us, would see all the galaxies (including ours) rushing away from them. The more distant the galaxy, the more rapid its recession from the observer.

In other words, the universe seemed to be expanding. Now, even *this* was an observational fact that cosmologists could use for scientific show and tell. What's more, it was a piece of observational data that fit previous theoretical ponderings.

Hubble solved astronomy's distance problem with the cosmic yardsticks offered by Cepheid variable stars. Along with his associate Milton Humason, Hubble measured the distances to more than forty galaxies whose speed of recession (or radial velocities) had first been calculated by Slipher from their spectral redshifts. Hubble's observations and calculations revealed a remarkable correlation between their distances and their velocities: The greater the distances, the larger their velocities. Some of the galaxies he observed had recessional velocities approaching 2,000 kilometers per second. It added up to a preliminary but convincing case for a relationship between redshifts and distances from the earth.

In 1931 Hubble and Humason published a second paper containing the results of more galactic observations, combined with those of Slipher from nineteen years earlier. Again, the distances and velocities of the galaxies were in direct proportion. Hubble and Humason had detected direct evidence for the expansion of the universe. The relationship between the velocities and distances of galaxies is known as the Hubble law. The rate at which the velocity of recession of distant galaxies increases with their distance is the Hubble constant.

Just as the cosmological principle applies to the universe only at its largest scale, so its uniform expansion applies only to the universe at large. The solar system, the galaxy, the Local Group, the Coma-Sculptor Cloud, and even the Local Supercluster are not expanding from one another. The mutual gravitational interaction of their constituent parts—planets, stars, and galaxies—holds them together.

The uniform expansion of the universe observed by Hubble and Humason supported the validity of the cosmological principle. It also provided cosmologists with something solid upon which to base their cosmological models. With Hubble and Humason's work cosmology had in fact (if not in the minds of some scientists) become an observational science. Any model for the origin of the universe had to account for the observed facts, including:

- The various levels of large-scale structure in the universe.
- The greater overall homogeneous distribution of matter in the universe.
- The uniform expansion of the universe.

Cosmological models also had to take into account the equations and consequences of general relativity, including the curvature of space-time caused by the presence of matter. As cosmologists explored the first few micro-instants of the universe's existence, they would also have to include the powerful implications of the newly emerging discipline of quantum mechanics. The next six decades would prove to be fruitful ones for cosmology and cosmologists.

The Big Bang

The observations by Hubble and Humason were pretty convincing, but they raised a question: What started the expansion? It was the kind of question that bothered plenty of scientists. One was Sir Arthur Eddington, who at the time was one of the world's best-known astrophysicists. Like others of his colleagues in astronomy, physics, and other disciplines, Eddington placed considerable weight on the aesthetics of a theory. Did it "feel" good? Did it seem to have beauty or symmetry to it?

Concern with aesthetics might remind us of Goethe, the German Romantic scientist-writer. As noted earlier, Goethe based his peculiar theory of light and colors on its aesthetics. It felt good to him and meshed with his convictions about the mystical powers of colors of light.

For Eddington, however, aesthetics was an essential part of good science. To him the most aesthetically pleasing cosmological theory was the one first proposed by Einstein—a static universe that was neither expanding nor contracting. Eddington proposed a way to reconcile a static universe with the undeniable redshifts found by Slipher, Hubble, and Humason. The universe, Eddington suggested, developed infinitely slowly from a primitive uniform distribution in unstable equilibrium. In other words, the universe was static but unstable. Tiny instabilities in this primitive state of being slowly built up and eventually caused the expansion observed today. In this way Eddington accepted the observed evidence for expansion, but avoided any definite moment of *creation* for the universe. Writing in 1932, Eddington said he wanted to avoid any requirement for "a sudden and peculiar beginning of things. I simply do not believe that the present order of things started off with a bang."[3] Prophetic words, indeed, but not in the way Eddington intended.

In the 1930s and 1940s two cosmologists, Georges Lemaître and the Russian-American physicist George Gamow, combined Friedmann's theorizing and Hubble's observations into a new theory. Several researchers, including Eddington, had been looking for possible explanations for the expansion observed by Hubble and company. Lemaître was the first to present a detailed suggestion of the simplest and most straightforward option: that the cosmic expansion then observed really did begin at the beginning of the universe. Lemaître looked at the evidence of a cosmological expansion of the universe, and essentially played the movie backward. What is large now must once have been small—very small. The farther back in time one goes, the smaller the universe must have been. The conclusion, for Lemaître, was obvious. In 1931 he suggested that all of the mass and energy of the early universe must have been concentrated in a "pre-universe" of matter and energy that he called "a primordial atom." This atom exploded, resulting in the creation of the universe.

Several years later George Gamow also proposed the "exploding primordial atom" concept as the origin of the universe. Gamow was an effective writer and popularizer of cosmology, astronomy, and physics. His books and articles spread the idea of "the primordial atom" among the general public, and he often gets credited with its invention. In the mid-1940s Gamow and two colleagues, Richard Alpher and George Hermann, took another step. They suggested that the explosion of the primordial atom would have left behind a "fossil" that would take the form of faint electromagnetic radiation in today's universe. In proposing this fossil, the three were offering a concrete observational test for the hypothesized beginning of the universe—and testing hypotheses is what science is all about.

In 1948 a competing cosmological model appeared on the scene. Proposed by Austrian astronomers Hermann Bondi and Thomas Gold and British astronomer Fred Hoyle, it was called "the Steady State" or "Continuous Creation" theory. This cosmological model proposed that the universe has no beginning and no end. It has always existed and will always exist. It is expanding because matter is being continuously created out of nothing at all, in the depths of intergalactic space. The exact mechanism for the continuous creation of matter was always a bit vague in this theory. The observational evidence to support it was less than vague: It was nonexistent. But it was a theory, proposed by three men with good credentials, and it was out there to be shot down or supported by whatever observations could be made.

The steady state theory relieved the nagging discomfort that many scientists had with any suggestion that the universe had a definite beginning. Beginnings imply creation . . . and a creator. Many if not most cosmologists did not wish to engage in unwinnable arguments with true believers about the existence or nonexistence of a Great Watchmaker in the Sky. If the universe had no beginning, then it had no creator. That solved *that* awkward problem.

The steady state theorists, who perhaps had read Eddington's book and his comments about the universe starting with a bang, scornfully referred to the Lemaître–Gamow original event as the "Big Bang." However, negative publicity doesn't always work in science. Gamow embraced the term Big Bang with delight and, turning the tables on the steady-staters, promoted the hell out of it. More importantly, observational evidence to support the steady state theory never materialized. Worse, evidence would soon be uncovered that would essentially shoot down the steady state theory for good.

Hearing the Echoes of Creation

Radio astronomy is the branch of astronomy that studies heavenly bodies by means of the radio waves they emit or reflect. It had its beginnings in 1931, when an engineer named Karl Jansky accidentally detected radio waves coming from the sky. In 1937 the American astronomer Grote Reber built the first radio telescope and began to systematically study the universe with it.

Radio telescopes act like giant "ears," "hearing" the radio waves that our eyes cannot see. When computers started being used to turn the data from radio telescopes into pictures, radio waves literally became visible. Radio telescopes detected exploding stars and colliding galaxies. The universe turned out to be stranger and more violent than anyone had imagined. Even more astonishing, with radio telescopes, astronomers actually detected the echoes of creation.

Thirty-six years after Hubble's discovery of the expanding universe, two researchers working for Bell Telephone Laboratories were testing some new and highly sensitive radio communications equipment attached to a large horn-shaped radio dish. As they did so, Arno A. Penzias and Robert W. Wilson were frustrated by some electromagnetic "noise" that was entering their antenna. The noise was in the microwave region of the spectrum, at a wavelength of about 7 centimeters. The intensity of the radiation at that wavelength was equivalent to that of a temperature of about 2.73 kelvins,[4] or 2.73 degrees Celsius above absolute zero. Wilson and Penzias mentioned their problem to Robert H. Dicke, a cosmologist at Princeton University. Dicke was intrigued by two researchers' findings. The Big Bang theory developed by Gamow and Lemaître offered a specific scientific prediction. An "echo" of that ultimate primordial event should still be present in today's universe in the form of "background" radiation permeating the cosmos.

The Big Bang theory is basically quite simple, but its implications are profound. As proposed by Gamow and later refined by others, the Big Bang theory states that the entire universe was once contained in a zero-dimensional point of infinitely high temperature and density. Out of this "singularity," as physicists describe it, came the entire universe. The Big Bang created more than all of the matter and energy in the universe; it also created space itself, and time.

In the first infinitesimal instants of its existence, all of matter and energy was one. The four basic forces we know today—gravitation, electromagnetism, strong nuclear force, and weak nuclear force—did not exist; only one force existed. As the universe expanded, it cooled. When steam cools, it becomes water. When water cools enough, it becomes ice. Chemists call these transformations changes of state. In an analogous fashion, as the universe cooled it also "changed state" and went through a process physicists call symmetry breaking. The force "cooled" and "broke" into two forces . . . then three . . . and finally the four we know today. Meanwhile, the infant universe cooled enough for quarks and electrons, the basic building blocks of matter, to "condense" into existence. The quarks quickly hooked up to create protons and neutrons. However, the universe was still incredibly hot, filled with a dense soup of radiation and charged subatomic particles such as protons and electrons. Matter and radiation were joined together in a close embrace—so close that the Universe was optically opaque. Photons, the particles that form light, were not able to move very far without hitting and interacting with free-flying electrons.

This state lasted until the temperature of the universe dropped to about 5,000 kelvins, about 300,000 years after the Big Bang. At that point the universe was cool enough for protons and electrons to combine and form atoms of electrically neutral hydrogen. Very quickly, the universe became transparent, since hydrogen atoms do not easily interact with photons of light; matter and radiation went their separate ways. The famous Biblical phrase from Genesis, "Let there be light," is therefore scientifically accurate. It just happened to be free to travel across the cosmos thirty thousand centuries after the Creation. This light is the fossil that Gamow and his colleagues had predicted would be detected if their Big Bang theory was correct. In the billions of years since that moment of matter–radiation decoupling, the "light" from the Big Bang has progressively gotten cooler.

P. James E. Peebles, a researcher in Dicke's group at Princeton University, carried out some calculations to determine the present temperature of that radiation fossil from the beginning of space and time. His answer: about 3 kelvins. The electromagnetic "noise" that Wilson and Penzias had found was microwave radiation at a temperature that matched the theory's prediction. Not only that, it was isotropic, or uniform, in every direction. No matter at what part of the sky Wilson and Penzias pointed the antenna, they found the 2.73-kelvin radiation signature. To the limits of their instrumentation, they could find no variation in this temperature, which suggested that the origin of the background radiation was extremely compact and uniform in nature. It was just what one might expect from the sudden expansion—from nothing at all, it seemed—of all space, time, matter, and energy.

Robert Wilson was not convinced, at first. He and Penzias looked for other possible causes of the 2.73-kelvin background radiation. They even investigated the possibility that pigeon droppings in the giant radio antenna horn could be the cause of the radiation signature. However, pigeon poop—along with every other possible explanation—lost out to cosmology. Other researchers followed up on Penzias and Wilson's discovery, and observed the background radiation at other microwave frequencies. High-altitude balloon observations in 1975 supported the blackbody nature of the background radiation. A *blackbody* is an object that absorbs all wavelengths of radiation falling on it, reflecting none of it back into space. A blackbody also emits radiation with perfect efficiency. In 1989 a satellite named COBE (Cosmic Background Explorer) decisively confirmed it. It was indeed the echo of the universe's birth wail. It reverberates still, across a cosmos now grown to at least 15 billion light-years in radius. The universe is not eternal; it had a beginning.

One of the most delightful, provocative, and instructive popular science books of this century is *Flatland*, by Edwin Abbott. In this "mathematical fantasy" (as the author termed it), fictional creatures living in the two-dimensional world of a plane confront the mathematics of a third dimension as well as the frustrations of trying to imagine the shape and feel of a world composed of three dimensions instead of the normal two. They cannot perceive the third dimension of height/depth with their senses. They have evolved in a three-dimensional universe, but their species' survival does not require the perception of the third dimension and so their visual cortices do not possess the ability to recognize it. In effect, they cannot look "up" or "down." If a three-dimensional being with two feet were to walk over and stand in their land, the flatlanders would see only two two-dimensional objects with the silhouettes of footprints.

Today, we face the same difficulties in imagining what the Big Bang was "really like." We have a picture in our minds. It is synthesized from our own limited experience with expanding objects (such as balloons) and pictures we have seen (*Dr. Strangelove*'s exploding H-bombs, a computer-generated Big Bang on *Nova*). The images we conjure up are three-dimensional, however. We do not and cannot perceive space-time as a unified entity. We perceive the three dimensions of space, but we cannot perceive time as a fourth dimension. Time is an arrow running from past through present to future, at a rate of one second per second. We are always in the *now*. The image we create of the Big Bang, the picture we see in our mind's eye, is of an explosion of all matter and energy sweeping out from a point and expanding forever.

But the Big Bang was not that at all. The universe is expanding, without doubt. What we find difficult to grasp, however, is that space-time itself began with the Big Bang. The universe is expanding, and that expansion is only sort of like an expanding balloon or an explosion. Balloons not only have space within them, but also space outside them—the space into which they are expanding. Explosions take place underground, undersea, in the air, in the vacuum of space—but they take place some *place*.

The Big Bang did not occur "some place." With apologies to Gertrude Stein, there is no "out there" out there. There is no way to speak in a scientifically rational way about "what lies outside the universe." Such a phrase is a philosophical contradiction in terms. "Outside" and "inside" are characteristics inextricably connected to space. In somewhat the same way, "past," "present" and "future" are inherent characteristics of time. Space is a property of the universe and exists only and wholly in the universe. Nothing exists outside the universe, because the universe is all that is. Time, too, is a property of the universe, and so there is no past or future time beyond the universe, before the universe began, or after the universe ends (if it ever does). Space-time itself begins with the beginning of the universe.

The Inflationary Vision

Any cosmological model that seeks to explain accurately the creation of the universe has to explain the constant rain of microwave radiation that falls through the cosmos. The Big Bang theory not only successfully predicted its existence, but also explains why it must exist. However, a successful cosmological model also has some other explaining to do. Wilson and Penzias's discovery of the cosmic background radiation, first predicted more than a decade earlier by Gamow and his colleagues, only served to raise more questions. The most important, and most vexing, was, If the Big Bang was as smooth an event as the uniformity of the cosmic background radiation implies, where did all the structure come from? The simplest version of the Big Bang theory predicts that a hot early universe would produce a radiation fossil with a smooth blackbody curve. That's what Penzias and Wilson had discovered in 1965. At the same time, a universe that began with a completely smooth Big Bang would still be smooth. It would contain no structures, but would be filled with an evenly distributed amount of matter and energy. However, we live in a cosmos that is far from smooth. The universe contains a myriad of structures, on scales ranging from the ultramicroscopic to the gargantuan. In the last several decades, astronomers have begun to understand just how gargantuan some of those structures are. As we've seen, the largest structures in the universe appear to be the galactic superclusters. Some of them stretch in sheetlike form for more than 500 million light-years. Indeed, astronomers have dubbed one such structure "The Great Wall."

But at the largest distance scales in the cosmos the cosmological principle rules. Distribution of matter in the universe is essentially homogeneous on scales exceeding several hundred million light-years. The Great Wall's length is just three-hundredths of 15 billion light-years, the effective radius of the universe. It's important to remember, of course, that the universe doesn't really have an "edge" to it. Our picture of it as a sphere or a balloon is totally misleading in this respect. The edge of the universe arises from our own sensory limitations, and from a basic law of nature. Much of our knowledge of the world comes from our sense of vision, our ability to perceive a narrow range of electromagnetic radiation. Visible light, and all electromagnetic radiation, travels at a maximum speed of 299,924.58 kilometers per second. So the edge of the universe, as far as we or any other light-perceiving being is concerned, is a "light horizon." That horizon lies 15 billion light-years away. We can never see any object further from us than that. Its light would take longer to reach us than the universe is old.

And that's where the structure problem lay. The time it would take for light to travel from a point near the universe's light horizon on one side of the sky to a point near the light horizon on the other side of the sky is greater than the age of the universe. This means that regions lying near the light horizon on opposite sides of the sky have apparently never been causally connected to one another. And here is the rub: With no causal connection between the opposite regions of the universe, no physical process could possibly have caused these regions to have exactly the same temperature and density at the beginning of the universe. So why do these regions appear to have exactly the same temperature for their cosmic background radiation? The only possible explanation is that the Big Bang was incredibly smooth. Absolutely no jerks, hiccups, or micropauses allowed. But . . . a Big Bang that smooth does not allow the conditions for creation of *any* structure in the universe. If the Big Bang had actually happened this smoothly, the universe would contain either matter and energy combined in a perfect solution—or nothing at all. Cosmologists thus found themselves caught between the Scylla of the night sky and the Charybdis of their equations. It was a decidedly uncomfortable place to sit.

In 1980 a young physicist dreamed up a daring solution to the Big Bang's causality conundrum. His name is Alan Guth, and he called his revolutionary cosmological model "inflation," or the inflationary universe. Guth suggested that mere instants after its creation, the universe underwent a superfast expansion. During this very short period of time, the universe's size increased exponentially. That is, during any specific time interval, the universe became twice as large as it became during the previous interval.

The cosmos was only about 10^{-35} seconds old when it began inflating. That's ten trillion trillion trillionths of a second After the Beginning (A.B.). Every 10^{-35} seconds the universe doubled in size, until it was 10^{50} times bigger than when inflation began. Guth's inflation equations suggested that when the inflationary period ended at 10^{-32} seconds A.B., the universe had a light horizon radius of about 10^{-25} centimeters. For comparison, the average width of a human hair is about 100 microns, or 10^{-2} centimeters. The radius of a hydrogen atom is a million times smaller, about 10^{-8} centimeters. A proton is ten thousand times smaller still, with a radius of about 10^{-13} centimeters, but is still a *trillion times larger* than the universe was at the end of the inflationary period.

The standard Big Bang theory presents a universe that has been expanding since the beginning. The expansion has been steadily slowed down by the gravitational force of all the matter and energy within it. Using the standard Big Bang model to "wind back" from the universe we see today to 10^{-32} seconds A.B., we end up with a universe about a millimeter in radius. That's 10^{24} times bigger than its light horizon at that moment. But the inflationary scenario predicts that today's observed universe was then about 10^{-26} centimeters in size.

What's going on here? Two things, actually—one of them good for cosmology and the other a godsend for science fiction writers. First, inflation solves cosmology's causality problem. A universe that's 10^{-26} centimeters in size is ten times smaller than its maximum light horizon region of 10^{-25} centimeters. The causality problem disappears. Widely separated regions of the cosmos lying at the edge of today's light horizon were in fact causally connected during the first several 10^{-35} seconds of the Universe's existence. At the same time, any initial lumps, bumps, burps, or jerks in the universe's initial expansion from nothing—earlier than 10^{-35} seconds—would have been utterly erased by this extraordinary growth in size. The universe's uniformity on the largest distance scales thus makes perfect sense, and is neatly and simply explained by inflation.

The second consequence is considerably more bizarre. Guth's equations conclude that when the inflation process began at 10^{-35} seconds A.B., the universe we finally see today became divided into 10^{72} regions with their own light horizons. We cannot see any of the other regions. They lie "on the other side" of the light horizon, and are therefore forever beyond our ken. We cannot act on any object in those regions, nor can any physical force from them reach us. And we cannot communicate with those other regions, nor they with us, for information can only travel at the speed of light. These other $10^{72} - 1$ regions are completely independent of us. They go their own way and we go ours. This is a real-life scientific explanation for an old science fiction device—the "multiple universes" concept. In science fiction stories, however, characters travel to and from alternative universes at the drop of a hat. In the "real universe," such travel is forbidden. The borders are closed and you can't get a passport.

Visions of Quantum Cosmology

Guth's inflationary model also offers answers to two other big questions left unanswered by the standard Big Bang model:

- Where did all the matter and energy in the universe come from?
- Where did the universe come from in the first place?

The answers are, briefly, supercooling and the free lunch program.

These answers flow naturally from the solution to another question: What caused inflation? The answer Guth provided comes from quantum mechanics and quantum cosmology, a realm of very small sizes and moments of time. So far we've been throwing around some pretty tiny numbers describing some extremely small distances and slices of time. Now we'll get even tinier. First, we look at the effects of gravity at the beginning of it all.

Of the four fundamental forces in the universe—strong nuclear, weak nuclear, electromagnetic, and gravitational—gravitation is the weakest. Although it is a long-range force, which becomes manifest when matter clumps together in large masses (gravity's web reaches to the farthest reaches of the cosmos), gravitation has just 10^{-39} the strength of the strong nuclear force. At the scale of atoms or protons, gravity is much too weak to compete with the other three forces. Gravity rears its head again at extremely small size scales, at distances less than 10^{-33} centimeters—10^{-20}, or a hundred billion, billion times smaller than a proton's radius. Physicists call this distance the Planck length. Along with the Planck length comes a unit of time called the Planck time, the time it would take for light to travel a Planck length. The Planck time is about 10^{-43} seconds. Planck time is the fundamental unit of time, the smallest tick of any clock. And we can think of the Planck length as the smallest notch on any ruler.

Next, there's the matter of uncertainty and the vacuum. As we'll see in more detail later, one of the most important bricks in the building called quantum mechanics is the uncertainty principle, first formulated by German physicist Werner Heisenberg. In the quantum reality world of atoms and smaller entities, certain values can never be known with certainty. For example, when we look at an electron that is part of an atom, we can measure certain properties, such as its position and its momentum. However, the more precise our measurements are of its position, the more uncertain is its momentum. And vice versa. Our attempt to measure the electron's position and momentum contributes to, but is not the fundamental cause of, this uncertainty. Rather, the uncertainty is a fundamental property of the quantum realm of reality.

The uncertainty principle is also at work in the apparent vacuum of "empty space," which is actually laced with electromagnetic and gravitational fields. The classical Newtonian view had been that empty space was just that. But quantum mechanics, with its uncertainty principle, basically forbids the existence of "nothing." "Nothing" is too definite a property, and the uncertainty principle says, "Wait a minute. Are you *sure*?" Indeed, we are not. Even in the "emptiest" vacuum, an electromagnetic or gravitational field exists that (1) has some measurable value or strength that (2) changes with time, and the uncertainty principle applies to those two properties. Some fundamental uncertainty will always exist about one or the other. According to quantum mechanics, then, a quantum vacuum is not really "empty." It is merely the lowest energy state possible for the entire system contained within it and that contains it.

This basic uncertainty about the energy of the vacuum can be thought of as quantum fluctuations in the vacuum. These fluctuations can be described as the appearance and disappearance of pairs of oppositely charged subatomic particles. One particle in the pair is normal matter; the other is a particle of antimatter. These "virtual particles" are the same as those Richard Feynman invoked in his description of quantum electrodynamics, or QED. They are temporary, and thus "virtual." The universe doesn't care that matter and antimatter have suddenly appeared out of nothing. That's because almost as soon as they appear, they disappear. The two particles in the pair meet and annihilate one another. We cannot directly measure their presence the way we can measure the presence of a proton. But we can indirectly sense their presence by their effects on other entities. For example, physicists have measured changes in the energy of electron orbits in atoms caused by virtual particles popping out of the quantum vacuum.

Quantum physicists have long known about the quantum nature of the vacuum and the existence of virtual particles. In fact, one way to describe the action of the four fundamental forces of nature is with virtual particles. These kinds of virtual particles are "carriers of the force." The carrier particle for the electromagnetic force is the photon, the quantum of light. The strong force's carrier particle is the gluon. The weak force has three carrier particles, the W^+ and W^- bosons and the Z^0 boson. Scientists have (indirectly but conclusively) detected the presence of all these virtual particles. Finally, the gravitational force can also be thought of as having a carrier particle: the graviton. Physicists have not detected gravitons yet.

To describe how the universe's inflationary period began—and ended, for that matter—Alan Guth turned to quantum mechanics and its description of the quantum vacuum. The equations of quantum mechanics describe a special state of the vacuum—an "excited" or "false vacuum" that exists at a higher energy level than the standard vacuum. The vacuum can stay in this state only briefly; then it releases its energy, which turns into matter—real matter, not virtual particles. What's more, while it is in this excited state, the vacuum possesses a force best described as "negative pressure" or even "antigravity."

Einstein's general relativity theory reveals that pressure is a source of gravity, just as the presence of matter is. Ordinarily, we think of pressure as a force that "pushes out"; a negative pressure would therefore be a force that "pulls in." But general relativity's "negative pressure" is like "negative gravity." Guth's inflationary scenario suggested that the very early, very tiny universe existed in this false vacuum state. We'd expect an infant universe containing a false vacuum to be exerting negative pressure to collapse. Ah . . . but normal pressure is a source of gravity (says general relativity). Gravity is an attractive force. So negative pressure creates a kind of antigravity, exerting a powerful repulsive force—a force of expansion. Voilà: inflation.

Inflation started at about 10^{-35} seconds A.B. and ended at about 10^{-32} seconds A.B. because a false vacuum state is inherently unstable. Take a smoothly curved bowl (empty, please), turn it upside down, and place it on the floor. Now take a small ball—a ball bearing will do, for example—and place it carefully on top of the overturned bowl. The ball's position on the bowl is unstable. The smallest disturbance will send it rolling down the bowl's side onto the floor. Once it gets to the floor it will eventually stop. The floor is (presumably) the lowest point in your house. It's stable. This is basically what happened to the false vacuum in the early universe. When it decayed, two things took place. First, the powerful repulsive force disappeared, since no more negative pressure existed. Second, the energy imprisoned in the false vacuum had to go somewhere. Energy (once it exists) can be neither created nor destroyed; it simply changes form. And Einstein's famous equation from his special relativity theory, $E = mc^2$, tells us that energy can also turn into matter. The energy of the false vacuum turned into heat and matter. All the matter we see today—galaxies and stars, planets and people, water and air and earth—all came from the energy of the decaying false vacuum some 10^{-32} seconds after the appearance of the universe.

The Origin of the Background Echo

There is one fly in this ointment, though, and its name is antimatter.[5] When virtual particles appear and disappear, zipping out of the quantum foam that we call "empty space," they always appear in pairs. One particle is of ordinary matter and the other is its mirror image. The universe loves symmetry. It doesn't always insist on it, as we'll soon see, but it does have a definite predilection for balance. This preference applies to matter. Every basic building block of matter—electrons, quarks, and the particles made of quarks such as protons and neutrons—has its corresponding mirror image particle. The proton's antiparticle is called an antiproton. Its mass is exactly the same as that of a proton, but it has an opposite electrical charge. Thus the antiproton carries a negative, not a positive, electrical charge. The electron's antiparticle, the antielectron, or positron, has a positive electrical charge. Even electrically neutral neutrons have antineutrons (their spins are opposite). Protons and neutrons are made of combinations of the still-smaller subatomic particles called quarks. So antiprotons and antineutrons are made of combinations of antiquarks. Matter and antimatter are, for the most part, indistinguishable from one another. However, they do have a rather unpleasant reaction to one another when they come into contact. Their opposite charges and other characteristics cancel out. The result, as noted earlier, is that matter and antimatter annihilate one another when they come into contact, and turn into energy. That energy is in the form of energetic gamma rays.

Antimatter is not theoretical. It's real. Paul Dirac first predicted its existence in 1929. His prediction resulted from his successful reconciliation of Einstein's special theory with the early versions of quantum mechanics. Three years later American physicist Carl Anderson detected particles of antimatter passing through an experimental particle detector he had built at the California Institute of Technology in Pasadena, California. Anderson detected positrons, the antiparticles of electrons.

Figure 13. Caltech's Carl Anderson discovered the positron, and thus introduced first scientists and later the public to a new alternate reality: a universe with antimatter.

Today, astronomers and physicists have detected small concentrations of antimatter in various locations in the universe. However, it is clear that far less antimatter exists in the universe than one would expect from simple symmetry. Where is it? Where did it go? These questions become more pressing when considering the Big Bang theory. In the beginning, matter and antimatter must necessarily have formed in the universe in equal amounts. And if that was the case, in the extremely tiny early universe the matter and antimatter would have quickly come into contact. It should *all* have been annihilated, and we should not be here reading this book and asking these questions.

So it seems that the laws of physics are not completely symmetrical when it comes to matter and antimatter. The universe may like symmetry, but it likes matter better. In 1964, two American physicists named James Cronin and Val Fitch found something unusual about the behavior of a subatomic particle called the K^0 meson. The K^0 is an unstable particle that quickly decays into several other more stable subatomic species. Like other subatomic particles, the K^0 has its antimatter equivalent, the anti-K^0. Cronin and Fitch found that the K^0 and the anti-K^0 do not decay at exactly the same rates. The effect is small, but the consequences for cosmology are huge. When inflation ceased and the false vacuum energy converted into matter and antimatter, a small asymmetry developed. For every billion particles of antimatter created, there were a billion and one particles of matter. As the universe expanded and cooled, the matter and antimatter collided and annihilated one another. All that was left were the one-in-a-billion excess particles of matter—which became us and every other material object in the cosmos. The annihilations created a flux of gamma radiation throughout the early universe. It, too, is still with us. Because the universe has expanded so much in the last 15 billion years, the gamma radiation has cooled somewhat. Today it exists as ordinary heat radiation, at a temperature just 2.73 degrees Celsius above absolute zero—in other words, the cosmic microwave background radiation.

The Big Bang and Alan Guth's inflation variation thus explain quite well the creation of matter and energy in the universe, the almost total absence of antimatter, the overall expansion of the universe, and the existence of the 2.73-kelvin cosmic background radiation.

And as for where or how the universe itself came into being? If Guth is right, the universe and everything in it is a sort of "vacuum fluctuation" from nothingness. Universes pop into and out of existence all the time. Ours, however, "popped" into existence and kept expanding. First inflation and then the inexorable expansion of the Big Bang has blown the universe from nothingness into a cosmos 1.4×10^{23} kilometers wide and 15 billion years deep.

Poets use metaphor and allusion to intensify their verbal images of nature and personal experience. Scientists often use the same techniques when trying to explain their discoveries. When cosmologists say the 2.73-kelvin background radiation is an echo of the Big Bang and of creation itself, they follow in the footsteps of Wordsworth and Shelley, Keats and Byron—and all poets before and since the nineteenth-century Romantics.

Another way of imagining the Big Bang is to think of it as a fossil, a "remnant or trace of an organism of a past geologic age, such as a skeleton or leaf imprint, embedded and preserved in the earth's crust."[6] In somewhat the same way, the cosmic background radiation first detected by Penzias and Wilson is a remnant of a very distant age, embedded in space-time itself.

Cosmology, then, is like paleontology in this sense: Both sciences are involved in finding fossils. Cosmology studies the nature, origin, and evolution of the universe. The fossils that cosmologists search for are not the remains of ancient living creatures, but rather fragments left over from creation itself.

From our location in the temporal landscape of the late twentieth century it is difficult to appreciate that the discovery of paleontological fossils has changed the way we perceive reality. Yet it did. There are still people in Western society who believe that the earth was created a few thousand years ago by the Divine Word of God; that the fossils of dinosaurs, ancient plants, and even more primitive bacteria are somehow a "little joke" that God is playing on us; that the biblical creation stories are not myths containing truths but literal statements of historical fact.

Most of us, though, envision the world somewhat differently. We carry within us a view of reality that stretches into "deep time," vast temporal vistas whose existence has been made manifest by the stony images of ancient bones we call fossils. The earth, we know in our own bones, is *old*, around 4.5 billion years of age. We and the other creatures with whom we share this planet are only the latest in a wondrous caravan of life that began its journey over the changing landscape of Earth about 4 billion years ago. But this is not at all the vision of reality perceived by our ancestors of only a few hundred years ago—much less a few thousand. The deep time of dinosaurs and fossilized bacteria did not exist for them. Their reality was a small one compared to ours: a few thousand years deep, a few hundred kilometers high. Other humans in times past had other visions of reality: Cyclical movements of eternal time and infinite expanses of universes are not concepts unique to late-twentieth-century Western cosmology and quantum mechanics. But for us whose philosophical and spiritual background is Western Christianity and the ancient Greeks, this vision of reality we have today is a radical change from that of our ancestors.

Space-time came into existence some 15 billion or so years ago. Nearly a century of observational cosmology leads us to that inexorable conclusion. The vision of contemporary cosmology first began taking shape with Vesto Slipher's spectrographs of nearby galaxies and their redshifts; it became clearer with Edwin Hubble's masterful maps of the expanding universe; and it finally came into sharp focus with the discovery of a universal echo of the first moments of the universe. All these discoveries were made on the ground, at the bottom of Earth's ocean of atmosphere, by astronomers using ground-based optical and radio telescopes.

The fossils found in the ground of England, Germany, France, and the western wilds of America helped create a revolution in our perception of the reality of deep time. A similar kind of revolution in our worldview, about the reality of space-time and the origin of everything, is rolling toward us like a runaway train. The engine pulling that train is a satellite named COBE. This new tool of astronomers and cosmologists has uncovered new details about the beginning. COBE's findings dramatically support the prevailing scientific theory for the creation of the universe, the Big Bang. They also give strong support to the inflation scenario. They have opened up a new set of questions for cosmologists, astronomers, and physicists to happily delve into.

COBE has discovered the electromagnetic fossils from the very dawn of the universe. As the implications of paleontological fossils began to be understood and accepted by the average person in the street, the Western vision of reality shifted. In somewhat similar fashion, the cosmological findings from COBE will eventually reshape our vision of creation itself.

A Vision of COBE

The fossils from the dawn of creation found by COBE are akin to ripples in the cosmic background microwave radiation discovered by Arno Penzias and Robert Wilson. Those "ripples" held the key to solving a serious cosmological puzzle. Penzias and Wilson had found no directional variations in this radiation, nor had others who followed them. And that created a problem. The extreme smoothness of the background radiation, its isotropy as physicists would call it, suggested that the Big Bang had been extremely smooth in nature. But such smoothness and uniformity in the universe's first few instants could not have led to the obvious structure seen in the universe today. It was a major conundrum for cosmologists and astrophysicists, and a potential weak point for the Big Bang theory. In 1970 the first measurements of anisotropy (or variations) in the cosmic background radiation were reported. P. James E. Peebles, who in 1965 had helped Penzias and Wilson figure out the nature of their cosmic background radiation discovery, summarized those early findings in 1971. But the observations were new and rough, without a lot of detail. Cosmologists wanted more.

Peebles suggested a way out: precise observation. Newer and more sensitive instruments should be able to detect anisotropies in the cosmic microwave background radiation to a high degree of accuracy. A team of researchers led by the late Nobel prize-winning physicist Luis Alvarez decided to turn Peebles's suggestion into reality. In 1974 they put together a proposal for NASA to build a satellite that would probe the edges of space-time for any such variations in the cosmic background radiation. Alvarez later stepped away from what became the COBE project; he felt it was becoming too big. Alvarez favored a small, fast-response team of researchers working with a fairly simple experiment. Besides, being in his sixties, Alvarez felt he might not live to see a big project to its completion. (Sadly, Alvarez was right; he died in 1988, a year before COBE was launched.)[7]

At about the same time, another small team of researchers had also begun pitching a satellite project to explore the cosmic background radiation. This team of four people led by John Mather of NASA's Goddard Space Flight Center in Maryland included Mike Hauser, Rainer Weiss, and David Wilkinson. A third proposal came from a group of researchers from NASA's Jet Propulsion Laboratory (JPL) in Pasadena, California, that included Samuel Gulkis. A series of NASA decisions eventually led to the three teams becoming one. Led by Mather, the initial group included Hauser, Weiss, and Wilkinson, George Smoot from Alvarez's group, and Gulkis from the JPL group. John Mather was the primary author of the four-hundred-page formal proposal for COBE turned in on February 1, 1977. Not long afterward Mather, Hauser, and Weiss went to NASA headquarters, and Weiss made the pitch to the peer review committee. Then everyone waited for the decision: Would COBE beat out the many other proposals for new scientific satellites and get moved to the front of NASA's program queue, or would it have to wait at the back of the pack?

While NASA evaluated the proposal for a cosmic ray background explorer satellite, researchers were busy on another tack. A team including Rich Muller and George Smoot, both of whom worked with Luis Alvarez, decided to make observations of the cosmic background radiation using a U2 spy plane. For several years Smoot and Alvarez had been using high-altitude balloons to carry instrument packages into the upper reaches of the atmosphere, looking first for cosmic antimatter and later gathering data on the cosmic background radiation. Since NASA was already using modified U2s for scientific work, Muller and his colleagues put together an instrument and carried out the observations. Their work provided additional evidence that the cosmic background was not isotropic. One part of the sky was indeed slightly "hotter" than the other. This dipole anisotropy (*dipole* means "two poles") was not due to any truly cosmological bumps or clumps, since it was highly directional. Instead, one possible cause might be the rotation of the Galaxy itself. The region of the sky toward which the Sun and Earth were heading might tend to appear slightly warmer in cosmic microwaves, while the opposite direction might be slightly cooler. However, the "hot spot" was not in the direction of galactic rotation. The dipole anisotropy had been predicted to exist in the 1960s and was first observed in the early 1970s. A high-altitude balloon probe in 1979 would decisively confirm its existence and also find evidence of a quadrapole anisotropy. That experiment was led by David Wilkinson, and included Edward Cheng, who would also be a member of the COBE team.

By the end of 1977 Mather and his team had gotten the answer they had hoped for. The combination of Weiss's persuasive oral presentation, the written proposal, and much behind-the-scenes lobbying by Nancy Boggess, a NASA official who championed the project, had convinced NASA to formally approve COBE.

The satellite would carry three instrument packages to observe the cosmic microwave background radiation. The first instrument package was called FIRAS, for Far Infrared Absolute Spectrophotometer. Mather was the principal investigator (or PI), the person in charge of the instrument. The second was the Diffuse Infrared Background Experiment, or DIRBE. Mike Hauser was named the PI for this instrument. The third experiment was the DMR, consisting of six differential microwave radiometers. Smoot was eventually named its PI.

Many years would pass before COBE finally got into orbit. At first COBE was designed to be launched on a typical expendable booster rocket. Then NASA decided to use one of the new fleet of space shuttles to put COBE into orbit; the shuttle needed plenty of cargo on its manifests to justify its existence. COBE's managers had to redesign the satellite to fit into the Shuttle cargo bay. In the early 1980s, while COBE was being developed, Alan Guth proposed his inflationary scenario for the Big Bang. Other researchers came up with possible explanations for the missing mass, or cold dark matter, in the universe. If these theories were true, the fluctuations in the cosmic microwave background radiation would be smaller than the DMR as originally designed could detect. Smoot, David Wilkinson, Samuel Gulkis, and others had designed the original instrument. Now, in a series of steps led by various team members, the DMR was redesigned and its sensitivity greatly increased.

In January 1986 disaster struck. The Shuttle *Challenger* exploded during launch, killing the seven-person crew and forcing a halt to America's human-crewed flights into space. The COBE team had to redesign the satellite once more, this time to fit atop an expendable Delta booster rocket. COBE was delayed yet again.

Finally, in November 1989, COBE was successfully launched into a polar orbit from Vandenberg Air Force Base, on the coast of California just north of the town of Lompoc. It had been nearly fifteen years since John Mather had first proposed COBE. No backup satellite existed; if the Delta booster had exploded on launch, it would have been the end of the line.

Astronomers and cosmologists already work with two types of cosmic fossils. It is by examining these fossils and learning about their origins and nature that cosmologists have been able to confirm that the universe had a beginning some 15 billion or so years ago. One cosmic fossil we know about for sure is what we are made of. Nearly all the matter in the universe, more than 98 percent of it in fact, consists of the two simplest elements: hydrogen and helium. The hydrogen atom is simply a nucleus of one proton "circled" by one electron. Helium has a nucleus of two protons and two neutrons surrounded by a "cloud" of two electrons. Primordial hydrogen and helium were formed in the first one to three minutes of the universe's existence.

However, the matter in the universe has been changed through time. Gravitational forces clumped much of the primordial hydrogen and helium into the first generation of stars. Fusion reactions within those stars "cooked" hydrogen and helium into heavier elements. Many of the elements found on Earth today, from lithium and boron through carbon and silicon, were created in the thermonuclear cauldrons within the first generation of stars and their successors. Many of those first ancient stars exploded as supernovae at the ends of their lives. Other heavier elements, from iron and zinc all the way to uranium, were synthesized in those titanic explosions and in the supernovae explosions of later generations of stars. The explosions of the primordial supernovae scattered the synthesized elements across space. The same process happened in second- and third-generation star populations throughout the cosmos, and will happen again and again.

Another cosmic fossil is even older than the primordial hydrogen and helium. This is the cosmic microwave background radiation first predicted by George Gamow and discovered by Penzias and Wilson. This "relic radiation," as some cosmologists call it, dates to a time earlier than the first minute or so in the existence of the universe.

Cosmologists continue to search for other cosmic fossils. One that has not yet been detected is the gravitational background radiation. Another is neutrinos from the beginning of the Big Bang. The universe should have a neutrino background radiation, which, according to George Smoot, would be a neutrino flux of about one thousand per liter. Neutrinos are subatomic particles that have either very little mass or (like photons) no mass, which means that the force of gravity has little or no effect on them. They also have no electric charge, and thus are not affected by the strong nuclear forces that govern activities within an atomic nucleus. The only fundamental cosmic force that acts upon neutrinos, then, is the mysterious weak nuclear force, which governs certain kinds of subatomic decay processes.

A Vision of the Beginning

On April 23, 1992, at a meeting of the American Physical Society (APS) in Washington, DC., the COBE team announced COBE's latest findings. George Smoot, the PI for the DMR instrument, made two presentations. The first was to a relatively small number of scientists attending the APS meeting. The second, a formal press conference, was to a packed audience of journalists. It was almost exactly seventy-two years to the day since the "great debate" between Harlow Shapley and Heber Curtis. That public meeting had delineated the differences between two competing visions of the cosmos. The COBE press conference offered the latest cosmological vision. COBE's differential microwave radiometers, their sensitivity increased at the insistence of the team, had detected ripples in the cosmic microwave background radiation.

Edward L. Wright, a member of the DMR team, was the first to find evidence of the ripples in the COBE data at the end of summer 1991. Not the dipole anisotropy that cosmologists already knew about, these ripples appeared to confirm findings made ten years earlier of a quadrupole anisotropy. (Two teams had made this initial finding; one was led by David Wilkinson, and the other by Francesco Melchiorri of the University of Florence.) The dipole anisotropy was fairly easy to explain. It was caused by the movement of the earth through the cosmic background radiation. This caused one pole or region in the sky to be slightly warmer than average, and the opposite pole or region of the sky to be slightly cooler. But a quadrupole is four poles: two warm poles and two cool ones. If this COBE finding was in fact correct, it would be the first indication of anisotropy in the cosmic microwave background radiation that was not related to the motion of the earth or of the Milky Way Galaxy.

As Wright analyzed the data, the more convinced he became that they showed evidence of a quadrupole anisotropy. Then other members of the team became convinced, including Charles Bennett. A researcher at the Goddard Space Flight Center, Bennett was Smoot's deputy PI. Smoot, however, appeared to remain skeptical about Wright's findings. In the midst of this intense puzzle-solving effort, Smoot left for the South Pole and a long-planned series of measurements of galactic radiation at wavelengths not covered by COBE. The other team members found this to be a surprising move. By the time he returned, Bennett, Wright, and team member Alan Kogut had all become convinced that COBE had indeed found indisputable evidence of a quadrupole anisotropy. They had written draft scientific papers, but Smoot—for various reasons—did not favor immediate publication.[8] Wright and Bennett wanted to go public with their discovery at the January 1992 meeting of the American Astronomical Society (AAS). Other team members did not. David Wilkinson, for example, wanted to make absolutely sure that any error in the data was essentially negligible. The discovery, if true, was so important that he and other team members wanted every "t" crossed and every "i" dotted.

By the time the team had done so, it was too late to make the announcement at the AAS meeting. However, they had confirmed beyond a doubt that COBE had found a quadrupole anisotropy. The DMRs had been pressed to the limits of their precision. The space-time ripples they detected were changes in temperature of less than 30 millionths of a degree from the overall 2.73-Kelvin background temperature of the cosmos, or one part in a hundred thousand of the total signal. During the first year of COBE's life, the six radiometers collected a total of about 400 million measurements of the cosmic background radiation's strength. Alan Kogut later described these measurements as pieces of a gargantuan jigsaw puzzle, a much larger version of the thousand-piece puzzles my father, brothers, and I used to enjoy working on. Of course, we had the picture on the front of the box to tell us what the completed puzzle would look like.

Suppose, though, that you had no idea what the final picture would look like. That's the situation in which Smoot, Kogut, Wright, Bennett, and the other DMR team members found themselves. Only when they found a way to fit all the half-billion measurements together did the picture begin to emerge. To do that, they used computer programs to statistically analyze the measurement data from COBE's differential radiometers—essentially a series of numbers. The numbers represented all the temperature readings of the cosmic microwave background radiation taken by COBE's equipment.

The statistical analysis did not make it possible for Edward Wright, or anyone else, to point to a place in the sky and say, "There—*that* is a ripple in space-time." It did, however, make it possible for the team members to extract meaningful information out of the mass of data. People generally have a difficult time digesting the meaning and implications of rows and columns of numbers. This is where computer and image-processing computer programs come in handy. The COBE team's computers turned the numbers into visual images in the form of colorful "maps."

"Most people relate to visual images," says George Smoot in an interview. "The data we gathered from COBE have a lot of noise in them." The "noise" Smoot refers to includes temperature readings from sources other than the Big Bang, for example. That noise or accumulation of meaningless numbers clutters up the collection of "real" data. "But the human eye," adds Smoot, "is very good at pattern recognition. For example, suppose you take a picture of a human face, like one formed on a TV screen from dots or picture elements, or pixels. You can put noise—extraneous pixels—in the image and you will still recognize the face.

"Now, compare that to the data we collected from COBE. The data don't stand out much above the noise. But we can still recognize a pattern. What we have is like a scrambled TV signal; you can see the faces, they're there all right, but you can't see much else. . . .

"So at this point we can say with strong confidence that *this* is noise and *that* is a real signal."

And what, exactly, has COBE been mapping? What do the wildly colorful splotches of color on the map really mean? What is the territory represented by the COBE map? According to Smoot, the COBE team has been creating a map of the signal caused by the curvature of space-time itself at the beginning of the universe.

"Think of space-time as a grid of lines drawn on the floor of a house," says Smoot. "Suppose there's an earthquake, and the house heaves around a bit, and the floor is no longer flat. Now it has curves or ripples in it. I remember once seeing an old Roman mosaic floor that was warped and rippled. Well, that's what we are seeing [in COBE's data]. Now, the main source of that curvature in space-time is the warping caused by the presence of matter/energy." This, of course, is exactly what Einstein was talking about back in 1916 in his general theory of relativity. General relativity is a theory of gravity—a geometrical theory of gravity. According to Einstein and his equations, space-time is a like a flat sheet of rubber. Take a ball bearing and place it on the sheet, and it makes an indentation. That dimple in the rubber sheet is gravity.

"What we are seeing in the COBE data," Smoot explains, "and what we're making a map of, are the impressions on space-time left there by matter in the very early universe. . . . The time scale we're working with is about 300,000 years after the beginning. That's how far back in time COBE looks. But we can already estimate how old the universe is," he adds, "which is somewhere between 12 to 20 billion years old. So if we see some evidence of a structure much bigger than 300,000 light-years across, then we know that the structure has to be older than that."

When we think about this for a moment, it makes simple sense. Any structure that exists in the universe when the universe is already 300,000 years old must have formed much earlier than 300,000 years after the beginning. The patterns of temperature differences found by COBE already permeated the early universe. They were essentially 300,000 or more light-years across. And that means they must have come into existence at a point *very early* in the universe's existence.

Says Smoot, "Quantum cosmology deals with events that occurred about 10^{-30} seconds after the beginning. And in fact, with COBE we are seeing stuff one to one hundred times bigger than that estimate I just gave you. So with COBE we are seeing structures that must date from very first instants of space and time."

Of course, the COBE results cannot stand on their own. They need to be corroborated by other scientists making other observations of the cosmic background radiation. As sure as the COBE team members have been about the validity of their conclusions, they also knew they could somehow have made a mistake. They anxiously awaited confirmation from other observers.

Nearly a year later they got it. In December 1992 a team of researchers from MIT and Princeton University announced partial confirmation of COBE's findings. In 1989 the team had launched a high-altitude balloon in New Mexico that carried an instrument to look for subtle variations in the background radiation. Like the COBE team, they had begun turning up cosmic fluctuations in their data. And like the COBE team, they spent months making sure that their findings were real. MIT's Stephan Meyer and Edward Cheng of the Goddard Space Flight Center (both COBE team members) and Princeton University's Lyman Page and Kenneth Ganga presented a map at a workshop held at the University of California at Berkeley showing fluctuations matching those found by COBE. In the years since, many other experiments both from the ground and from balloon-borne instruments have confirmed COBE's discovery of a quadrupole anisotropy.

The Meaning of the Visions

Nothing stays constant for long in the roiling atmosphere of today's cosmology. That applies to COBE's findings and the interpretations offered by Smoot and his colleagues. At the end of June 1992, cosmologists and astrophysicists met at Princeton University to hash over those findings and interpretations. The observations themselves went mostly unchallenged. The interpretations did not.

At their April 1992 press conference and in an earlier scientific paper, the COBE team had presented several interpretations and claims for the significance of the data:

- COBE's findings left no doubt that the Big Bang was the only serious scientific cosmological model. As Edward

Wright later said on CNN, "The Big Bang is alive and well."[9]

- The findings supported the inflationary variation of the Big Bang theory. The inflation scenario's calculations predicted almost exactly the kinds of temperature fluctuations seen by COBE.
- The data from COBE did not rule out the existence of so-called cold dark matter as the "missing mass" in the Universe.

For some time before the COBE press conference—and periodically thereafter, for that matter—the popular news media had trumpeted the assertion that "the Big Bang theory is in trouble," or that "the Big Bang is on the way out," or even that "the Big Bang is dead." This kind of reportage made for great headlines in tabloids, and even in otherwise respectable newspapers and magazines. In fact, the observational evidence for the Big Bang has been nearly rock solid since Penzias and Wilson's discovery of the cosmic microwave background radiation in 1964. Astrophysicists later measured the relative amounts of hydrogen, helium, and lithium in the cosmos and found that they matched almost perfectly the predictions from the Big Bang theory.

The DMR team also suggested that their results supported the inflation variation of the Big Bang theory. Inflation had been a major contender for the cosmological crown for nearly a decade. If an inflationary episode did in fact take place in the first 10^{-35} seconds or so of the universe's existence, expanding the size of the cosmos some 10^{50} times in the wink of an eye, it would have smoothed out any large "lumps" in the fabric of early space-time. Guth's inflation scenario predicted that the only thing left would be very tiny wrinkles in today's cosmic background radiation. And COBE saw just the kind of pattern in the cosmic background radiation predicted by inflationary models. The pattern, called a "scale-invariant power spectrum," basically amounts to spots within spots, similar patterns occurring at different scales.

Inflation is not the only model for the state of the very early universe. Other cosmological models also exist, most notably the cosmic string or cosmic defect model. At a Princeton meeting in June 1992, cosmologist David Spergel stood up in the audience and proclaimed that defect theories were dead. Charles Bennett simply stated that the COBE results were consistent with the predictions of inflation. And Neil Turok, a cosmologist from Princeton and proponent of the cosmic defect scenarios, noted that he felt COBE's observed power spectrum supported the inflationary model—and that it also agreed with the predictions made by cosmic defect models, as well.

Then there's the matter of the universe's missing or dark matter. Inflationary theory predicts that dark matter types of subatomic particles, massive and slow moving (hence "cold"), would be the "seeds" for the formation of galaxies and other cosmic structures. An alternative to the cold dark matter theory is hot dark matter. Unlike cold dark matter, which is still purely hypothetical, at least one form of hot dark matter exists, namely, neutrinos, massless (or nearly massless) subatomic particles that move at close to the speed of light. Other models call for mixtures of cold and hot dark matter. The data from COBE do not rule out any of the leading dark matter models, though they did put some constraints on certain combinations of parameters in these models. George Smoot felt that three-quarters of all the previously viable cosmological theories had been dealt a fatal or near-fatal blow by COBE. But that's Smoot's opinion, and other cosmologists strongly disagree—especially those whose models were supposedly assigned to death row.

COBE's findings also did not validate the so-called MaCHO model for the universe's missing mass. MaCHOs ("Massive Cosmic Halo Objects"), if they exist, are objects with one-hundredth to one-tenth the mass of the sun, often called "brown dwarfs" by astronomers. The MaCHO model suggests that the outer regions of galaxies such as our own, called galactic halo regions, contain untold billions of brown dwarfs and other MaCHOs. Their total combined mass might account for much if not most of the universe's missing mass. The problem—as always with cosmology—is a distinct dearth of observational evidence for the existence of MaCHOs. Speculation, as we've already seen, may be fun, but it's not science. In the case of MaCHOs, though, observation may be replacing speculation. At an astronomy conference in Italy in September 1993, two separate teams of astronomers announced preliminary evidence for the existence of MaCHOs in our own Milky Way's galactic halo.[10] However, several more years of observations now indicate that there is not enough mass in MaCHOs to have any significant cosmological effect.[11]

COBE's scientific discoveries were momentous. But the single biggest quote to come out of the April 1992 press conference didn't have to do with physics or cosmology. As "point man" for the COBE team at this public coming-out, George Smoot had performed brilliantly. His presentation was lively and engaging, and the media loved it. Toward the end of the proceedings, Smoot tried to put the significance of the COBE findings into a context that the average taxpayer could understand. He did so by using metaphor.

"If you're religious," said Smoot, "this finding is like seeing God."

Not surprisingly, the news media jumped all over Smoot's "God and religion" comment. The headlines screamed about scientists finding "God's fingerprints," detecting "the mind of God," and "uniting religion and science." "These results firm up one's view of the presence of some deity whose purpose is being worked out," said Arnold Wolfendale, Britain's Astronomer Royal in a Reuters news report. Then, perhaps in a belated attempt to cover his derriere, he cautiously added that "religion is a completely different dimension not susceptible to scientific proof."

Other DMR team members did not appreciate the "God and COBE" implications supposedly made by Smoot and trumpeted by the news media. One suggested that it gave people "the wrong idea about what it means to be a scientist." Another noted that "science is about things you can measure," and Divinity does not fall into that category. Donald Cupitt, a lecturer in the philosophy of religion at Cambridge University in Britain, also tried to put things in a different perspective. "Since Galileo . . . our whole vision of the world has come to be in terms of mathematics," he told Reuters. "[T]he old vision of the world . . . saw it as the expression of a personal purpose with moral values built in." Science, by contrast, with its positivist and materialist foundation, eschews any attempt to determine moral or ethical values. Added an editorial in the British newspaper the *Guardian*, "There is little chance of [COBE's findings] shedding ripples of light on the religion vs. science debate."[12]

Smoot himself was a bit taken aback by the furor. His comment about religion and COBE, he later said, was meant to put the importance of the COBE findings into a context easily understood by the general public. His comments were not meant to imply that COBE had found God's fingerprints, much less the Divinity. But words, once spoken, cannot be taken back. And some metaphors are so riveting that they take on a life of their own, despite the wishes of their creator.

The Alternate Reality: Cosmology

Cosmology is where all the realities we experience meet and exchange gossip among themselves. "It's kind of like metaphysics," Smoot remarks, "or what science used to be called, 'natural philosophy.' The meeting place is cosmology. It used to barely be considered a science, because it had such few facts to work with." In their 1996 book *The Very First Light*, John Mather and John Boslough quote a passage that Rainer Weiss wrote in the 1977 report to NASA for the proposed COBE project. "Cosmology, at present, is heavily weighted toward theoretical speculation because there is so little observational evidence."[13] Now cosmology is an observational science. And that in turn is leading scientists to ask new questions and to seek new methods and build new instruments to try and answer them.

Aesthetic judgment and "thinking"—call it speculating or even fantasizing if you want—can get scientists into trouble. Even making metaphors is dangerous at times, as Smoot himself discovered to his discomfort. Yet he remained adamant about the right and even necessity of scientists engaging in aesthetic speculations and philosophical ruminations: "People ask these questions! People want to know the origin of things, and the meaning of things. What is the purpose of it all? How does it all fit together? You try to explain what 'ten to the minus forty-second seconds' means, and you can't. And no one knows what that means. Cosmologists just know that these philosophical questions will come up. In cosmology, we are pushing the borders of science."

Smoot pauses for a moment, and then adds: "It's amazing, when you start thinking about it. You realize here we are on the earth and—it's amazing. You begin thinking about the sheer numbers of stars and planets, and you start getting some perspective. If inflation is right, then the entire universe we see is just a tiny speck [among 10^{72} universes]. You get a different perspective. The earth seems insignificant, but at the same time it also seems valuable and unique."

More than two centuries ago William Wordsworth fell in love with the landscape of England's Lake Country. As he walked through that countryside he experienced a flood of emotion; he would later recapture those feelings in powerful poems like "Tintern Abbey," poems that still move hearts today.

As beautiful as the Lake Country was, and still is, cosmology is the grandest of landscapes. It stretches from the beginning to the end of time and space. Mountains of "why" rise up from the plains of "how," and rivers of "where" lead forever on. We do not walk through it on foot, but traverse it entirely in our minds. We take the information gleaned from instruments like COBE and other satellites, as well as from observations made with radio telescopes, ground-based optical telescopes, and the Hubble Space Telescope, and sculpt it into the latest alternate vision of reality.

Yet even this vision will change, as cosmologists uncover still more secrets about the universe's beginning and end. In the next millennium we may well find ourselves living in a new, humbling, but also exalted, alternate cosmological reality.

Only one other alternate reality of science will have as powerful an influence on our vision of things invisible. And that is the alternate reality even now being created by quantum physics.

4

Quantum Realities

Someday perhaps the inner light will shine forth from us, and then we shall need no other light.

—JOHANN WOLFGANG VON GOETHE
Elective Affinities

The borough of Princeton, New Jersey, is home to one of the most prestigious educational facilities in the world. Princeton University was founded in 1746, thirty years before the beginning of America's war for independence. For more than fifty-five years now it has been the academic home of the man who today is arguably the most important figure in both physics and philosophy. His name is John Archibald Wheeler. He is a leader in the renewed dance of physics and philosophy.

Many people have contributed to this renewal. Most have been not philosophers but physicists. Alain Aspect, John Bell, Niels Bohr, Max Born, Hugh Everett III, Erwin Schrödinger, Werner Heisenberg, Albert Einstein, and Eugene Wigner (the last two were also Princeton residents) all played essential roles in the invention of quantum physics. They also tried to tackle some of the philosophical implications of their work. But Wheeler has not only made profound contributions to physics in several areas, but has also fearlessly ventured where angels and most other quantum physicists fear to tread—out to the borderland where physics and philosophy rub against one another like shifting continental plates.

Humanity got its first glimpse of what we could call "quantum reality" in 1900. Whether we regard that year as the last of the nineteenth century (as it is calendrically) or as the first of the

Figure 14. John Archibald Wheeler: master of spacetime and gravity, inventor of the term "black hole," and explorer of the observer-participant universe of quantum mechanics.

twentieth (as it is popularly accepted), it remains appropriate to refer to it as the opening year of the quantum era. John Wheeler would later write that "Planck's discovery of the quantum in 1900 drove a crack in the armor that still covers the deep and secret principle of existence. In the exploitation of that opening we are at the beginning, not the end."[1] He is profoundly correct in that assessment, as the philosophical implications of Planck's work continue to ripple through our culture.

We cannot perceive quantum reality with our unaided senses or with any possible form of instrumentation. The world of the quantum is preeminently the world of the probable and thus forever beyond our sense of physical sight because we cannot see probabilities, only actualities. The reality of the possible, of the "what if," of metaphor, is seen only with the inner eye or, if you're a physicist, with the "eyes" of mathematics. All things are possible at that level of reality, but few become actual. What makes them real is us.

The essentially mechanistic philosophy of positivism, which pervades modern science like the mythical ether supposedly filled all of space, arose in the eighteenth century as a direct result of the successes of the Newtonian vision of reality. Philosophers have no problem acknowledging those origins. Nor do they shy away today from the baffling and even disturbing philosophical consequences of quantum physics. Scientists, however, for the most part *have* shied away from considering the philosophical consequences of their work. Science as an overarching field of human activity has long tried to ignore its roots in the "natural philosophy" of earlier times. Centuries ago science and philosophy parted ways. Science insisted that its purview was "the real world," the world of things and actions and reactions and forces; the world of experiments and proofs and disproofs; the world of matter and energy. Philosophy sneered in response; let science muddy its hands in the seemingly solid world of matter. Philosophy would explore the world of ideas, using whatever ideas science might bring it.

Would that their ancient marriage had lasted. Under the historical circumstances, it would have required a miracle of counseling and therapy to bring them together. In our time, though, in our century, they are cautiously, reluctantly moving back toward one another. They dance again. And John Wheeler has been an eager participant.

Philosophical Visions

Quantum mechanics is called *mechanics* because it is a theory that tries to explain the movements and interactions of atoms and the even tinier entities of which atoms are made. Three hundred years ago Isaac Newton developed his theory to explain the observations of reality made by Galileo and others. The inventors of quantum mechanics fashioned their new theory for the same general reason: to explain scientific observations that Newtonian physics (classical mechanics, as physicists also refer to it) could not. When Newton's theories swept through the Western world, though, they seemed flawless.

In the seventeenth century, science as we know it today was just beginning to blossom. Galileo, Newton, Leibnitz, Hooke, Halley, and others were making discoveries that greatly expanded the Western world's understanding of the nature of the universe. At the same time, philosophers developed two different approaches to the search for knowledge. One was that knowledge could be obtained through reasoning, through the exercise of pure mental effort, a philosophical position called rationalism. The other was that observation and experience would bring knowledge, a position called empiricism.

The success of Newton's theories had led people in the eighteenth century to view the physical sciences as a model for all forms of knowledge. A mechanical philosophy of science and nature emerged during the period historians now usually call the Enlightenment. Chemistry and biology, as well as physics, it was believed, would ultimately turn out to be ruled by a set of simple mechanistic natural laws. Though at the time only partly successful in biology and chemistry, this mechanistic view of science was powerfully successful in physics and astronomy. The triumph of the mechanistic view in science was so overwhelming that it inevitably influenced the position of many philosophers, especially several influential French philosophers, including Denis Diderot. They denied the existence of God and saw nature as a completely mechanical, materialistic system. The laws of nature were reasonable laws, according to the major philosophers of the time. Nature itself embodied reason. They believed that history, social processes, and even spiritual yearnings could all be explained by a set of mechanistic natural laws.

In the latter part of the eighteenth century the German philosopher Immanuel Kant reconciled the positions of rationalism and empirisicm in his books *Critique of Pure Reason* (1781) and *Critique of Practical Reason* (1788). Kant pointed out that reasoning and experience go hand in hand. He also stressed the importance of causality, the idea of cause and effect. Kant said that causality was an a priori concept—a principle that cannot be proven, but must be accepted as true no matter what our senses might tell us. Other a priori concepts in Kant's view of the world were space and time.

At the same time that Kant, Diderot, and other philosophers were cementing the foundations of a materialistic view of reality, another philosophical idea was gaining prominence. Called the great chain of being, it had its roots in Aristotle and had been revived in the writings of Leibniz, the German philosopher and mathematician who along with Newton had invented calculus. This concept held that all of existence was continuous. For example, all animals were connected together in a continuous chain of existence from the tiniest mite to human beings. No gap existed among or between different types of animals. Eventually this idea helped give rise to the concept of "progress" as we accept it today. Over time, life progresses from primitive forms to more advanced forms. Over time, human society progresses from savagery to civilization. Just as some primitive forms of life still exist today, so some primitive and savage forms of humans still exist. It was the reasonable, inevitable, even supposedly "sacred" duty of the advanced white society of Europe to civilize the poor savages. And if they couldn't be civilized, then they should be eliminated.

The eighteenth century also saw a powerful reaction to this materialistic view of reality take place. In Germany it was called the school of *Naturaphilosophie*; in England and France, Romanticism. The German leaders included the poet-scientist Johann Wolfgang von Goethe and the biologist Lorenz Oken. In France the major proponent of Romanticism was Jean-Jacques Rousseau. In England the most prominent expression of Romanticism was in the poetry of the men we have encountered in this inquiry: Wordsworth, Coleridge, Keats, Shelley. The view of nature held by the Romantics was essentially a holistic one. Nature was all a single organism, and—just as importantly—all of nature was imbued with spirit.

The Romantic reaction to scientific materialism extended into the nineteenth century, supported in part by the writings of German philosopher Georg Wilhelm Friedrich Hegel. Hegel's philosophy of nature stressed the importance of intuiting the existence of a priori concepts over rationalistic experimentation. This position remained strong in Germany during the first part of the nineteenth century. Goethe also continued to be a force for the Romantic backlash against materialism. He tried to uncover scientific explanations by using general philosophical principles. One important example of this attempt was Goethe's opposition to Newton's theories of light. Newton, as we've seen, had used a prism to show that white light actually consists of a mixture of different colors. Goethe rejected this on philosophical and mystical grounds. He insisted that white light must be simpler and purer than any mixture of colored light.

Goethe's position seemed intuitively correct. But the prism doesn't lie. In the end the proponents of Romanticism and *Naturaphilosophie* could not stop the march of nineteenth-century science. It was just too successful.

Leading the intellectual parade was the French sociologist and philosopher Auguste Comte, born the same year (1798) that Wordsworth and Coleridge published the first edition of *Lyrical Ballads*. As noted earlier, Comte founded the school of philosophy known as positivism. He was influenced by the philosophical concepts of materialism, rationalism, and progress—which in turn owed their existence in large part to the steady successes of eighteenth-century science and technology. Comte proposed that humanity had progressed through three stages of social development. The first stage was theological: the belief that natural phenomena are explained by the existence of invisible and supernatural beings and forces. The second stage, said Comte, was metaphysical: the belief the natural phenomena can be explained by philosophical ideas, and that those ideas have an existence in reality. The third stage was positive: the realization that natural phenomena can be explained by the scientific process of observation, hypothesis, and experiment.

Positivism has had a profound effect on twentieth-century science. The usually unspoken assumption—the a priori concept, that is—in much of science in this century is that of positivism. The goal of knowledge is simply to describe the phenomena that scientists experience or observe. This is the ultimate materialistic variation on the Newtonian idea of a clockwork cosmos, and the culmination of three centuries of scientific revolution, a revolution that began with a small group of devout Christians who sought knowledge of the natural world only to further glorify the God who created it. From a Catholic canon named Copernicus to an atheistic philosopher named Comte is one hell of a jump.

Invisible Light, Invisible God

In 1800 Sir William Herschel, the discoverer of the planet Uranus, attempted to measure the temperature of different colors of light in the spectrum. In the process of his experiment he placed a thermometer below the reddest band of light. When he did so he found that the thermometer detected an invisible form of radiation that was hotter than visible light. Today we know this as infrared ("below red") radiation. The following year the German scientist Johann Ritter discovered a second form of invisible light. This radiation lay above the color violet in the spectrum and catalyzed chemical reactions in silver chloride more effectively than did visible light. We call it ultraviolet ("above violet") light, and it's the cause of that nasty sunburn we get at the beach on summer days when we don't smear on enough sunblock lotion.

The implications of these discoveries for what people knew about reality were profound. Humans have always believed that invisible beings and realms of reality exist. Our religions and spiritual philosophies are filled with stories of gods, goddesses, heavens, and hells. Now scientists had discovered something invisible that was *measurable*. God, angels, and the Devil were not detectable or measurable with scientific instruments; therefore (many educated people of the time concluded) they could not exist. Here, however, was an invisible reality that *was* measurable, that *did* exist. What other invisible realities would all-powerful science reveal to humanity?

By the middle of the nineteenth century science was well on its way to becoming a worldview with a profoundly materialistic and positivist philosophical foundation. At the same time, physicists were discovering more forms of light that humans could not perceive with their unaided senses. On the one hand, philosophers influenced by science and many scientists themselves denied the existence of any reality that could not be perceived. On the other, scientists were uncovering the existence of natural phenomena that only their instruments could detect. The invisible reality of gods and goddesses was being replaced by the invisible reality of ultraviolet and infrared light.

Wrench in the Works

The same year that Johann Ritter discovered ultraviolet light, an English physicist and physician named Thomas Young threw a monkey wrench into the Newtonian machine. Newton had postulated that light was made of tiny particles that he called corpuscles. Light was able to travel through space from, say, the Sun to Earth because it was made of these tiny particles. Newton's corpuscular theory of light made sense and it fit many of the observations made by physicists and others. It explained why mirrors worked, for example, and why white light passing through a prism broke up into its constituent rainbow of colors. Beginning in 1801, Young carried out a series of experiments that by 1803 showed that light sometimes acts like it is made not of particles but of waves. The kinds of experiments Young carried out are rather simple. Suppose we take two pieces of stiff cardboard-like paper. In the first sheet of paper we cut two narrow slits very close to one another. We position the second sheet of paper a short distance behind the first. Then we place a source of light in front of the first sheet of paper so that the light passes through the two slits and strikes the second sheet of paper. If light consists of tiny corpuscles we can confidently predict that we will see two slits of light on the second piece of paper.

That is not what we see, however, and it is not what Thomas Young saw, either. What we see on the second sheet of paper, and what Young saw, are a series of bands of light interspersed with bands of darkness. They are called interference patterns. Suppose we stand on the shore of a still lake and toss two small pebbles into it, one close to the other. Each pebble will kick up a series of concentric waves of water. If we watch closely, we will see that some of the waves from both pebbles cancel one another out, while others act to reinforce one another and create very tall wavecrests. The same type of phenomenon occurs when light passes through two thin slits placed very close to one another. We see interference patterns, and such patterns can only be created by the action of waves. Young was seeing waves of light interfering with one another. When the crest of one light wave hit the trough of another, they canceled each other out. The result: blackness. When the crest or trough of one light wave met the crest or trough of another, the two light waves reinforced one another and created a band of light.

Young was not the first scientist to discover the interference phenomenon of light. Francesco Grimaldi had noted it nearly two hundred years earlier. His speculations that light was made of waves appeared in a book published two years after his death. No one took it seriously, however.

Young at first took considerable heat for his discovery because it did not fit the prevailing paradigm. But the phenomenon he had discovered was real, and there was no denying it. Light did act like a wave. If light was a wave, moreover, then it must have a medium in which to move. Waves in water traveled in water. Sound was a well-known wave phenomenon that traveled through the medium of air. What did light move through? It must be something other than air, because light was certainly traveling from the Sun to the earth through the vacuum of space. Physicists concluded that all of space must be filled with an invisible substance that acted as the medium through which light waves traveled. They called it the "ether." The first scientist to apply this name to the space-permeating medium was the French physicist and engineer Augustin Jean Fresnel, taking the name from the fabled fifth essential element proposed by Aristotle. Fresnel's investigations of the polarization of light had led him to advocate the wave theory of light.

Within a few years Young's wave theory of light would gain powerful support from scientists working in two seemingly unrelated fields: electricity and magnetism.

Electrical Visions

Magnetism and static electricity have fascinated people since ancient times. The ancient Greeks were familiar with static electricity and knew about the magnetic properties of lodestone. The Chinese began using magnetic compasses for navigation around 1000 CE. The first Western account of the forces between the poles of a magnet and a compass dial appeared in 1269. In 1600 the English natural philosopher William Gilbert had published his book *De Magnete* (*On Magnetism*), in which he suggested that Earth was a giant spherical magnet. He also noted different substances that could be used to create static electricity.

During the seventeenth century comparatively little happened in the way of experiments with magnetism and electricity. But interest in both revived in the eighteenth century, as researchers and tinkerers began uncovering some of their secrets. In 1733 the French natural scientist Charles François Du Fay discovered that two types of static electrical charges existed. Objects with like charges repelled one another, while those with opposite charges attracted one another.

In 1746 Pieter van Musschenbroek invented the Leyden jar, the first practical way to actually store static electricity. Three years later an Englishman named John Canton developed a way to make artificial magnets. In 1750 the Reverend John Michell, an Anglican priest, explained magnetic induction in his publication *A Treatise on Artificial Magnets*. He also discovered the inverse-square law governing the repulsive forces of magnetism.[2] That same year Benjamin Franklin described electricity as a fluid and noted the distinction between positive and negative electricity. In 1751 Franklin carried out his famous kite experiment, showing that lightning was a form of electricity similar to the charge contained in Leiden jars. He also described how electricity could magnetize and demagnetize an iron needle.

Thirty years after Franklin flew his kite, a French scientist named Charles Augustin Coulomb undertook a careful study of the properties of static electricity and magnetism. In 1785 he carried out a series of precise measurements of the forces of attraction and repulsion between electrically charged bodies and between magnetic poles. He found that the strength of both the two electrical forces and the magnetic force obeyed an inverse-square law, rediscovering what John Michel had first uncovered, but not publicized, thirty years earlier. In 1786 Coulomb announced his results.

The inverse-square law states that the magnitude of a physical quantity is inversely proportional to the square of the distance from the source of that quantity. Suppose that object A is exerting a force on object B that obeys the inverse-square law. When the distance between them is 1 meter we measure the force and find that it equals 1 frappitz ("frappitz" means nothing at all; I've just invented a fictional name for the fictional units of this fictional force). If we increase the distance between object A and B to 2 meters, the strength of the force decreases. However, it does not drop to 0.5 frappitz, because the force does not follow a simple linear law. It follows an inverse-square law. So the strength of the force is equal to the *inverse* of the *square* of the distance; 2 squared is 4, and the inverse of 4 is $\frac{1}{4}$, or 0.25. The force between object A and B has dropped to 0.25 frappitz. When the distance between them increases to 3 meters, the strength of the force will drop to 0.1111 frappitz; at 4 meters it will be 0.0625 frappitz, and so on.

Newton's law of gravitation follows the inverse-square law. It states in part that the gravitational force between two bodies is proportional to the inverse square of the distance between the centers of mass of the two bodies. Coulomb's law is very similar. In its entirety, it states that the force between two point charges is proportional to the product of their respective charges as well as to the inverse square of the distance between them.

In 1791 the Italian physicist Luigi Galvani discovered that if a frog's legs were placed in contact with brass and iron at the same time its muscles twitched as if hit by a jolt of electricity from a Leyden jar. When another Italian scientist, Alessandro Volta, heard about Galvani's work, he quickly found that a simple chemical solution would replace the frog's muscles. By the end of the century he had invented the first electrical battery. A battery, unlike a Leyden jar, was a source of current electricity: electricity that flowed steadily through a conductor the way water flows in a stream.

The Connection between Light and Electricity

Something truly exciting was afoot. Coulomb had shown that both electricity and magnetism followed a mathematical law similar in nearly all respects to one that was crucial to Newton's famous gravitational force. Gravitation, however, was only attractive, while both magnetism and electricity could be attractive or repulsive. This implied a connection between electricity and magnetism. In 1807 the Danish chemist and physicist Hans Christian Oersted announced that he would find the connection. And he did—but by accident.

In 1819, twelve years after his bold announcement and eighteen years after Thomas Young's discovery of interference patterns of light, Oersted was conducting a classroom experiment. During the experiment he happened to place a compass needle next to a wire carrying an electric current. To his surprise, the needle of the compass jerked about and lined up perpendicular to the wire and the current running through it. An electrical current essentially turned the wire through which it was flowing into a temporary magnet (see Figure 15). The conclusion was inescapable; electricity and magnetism were not two forces but one; electromagnetism.

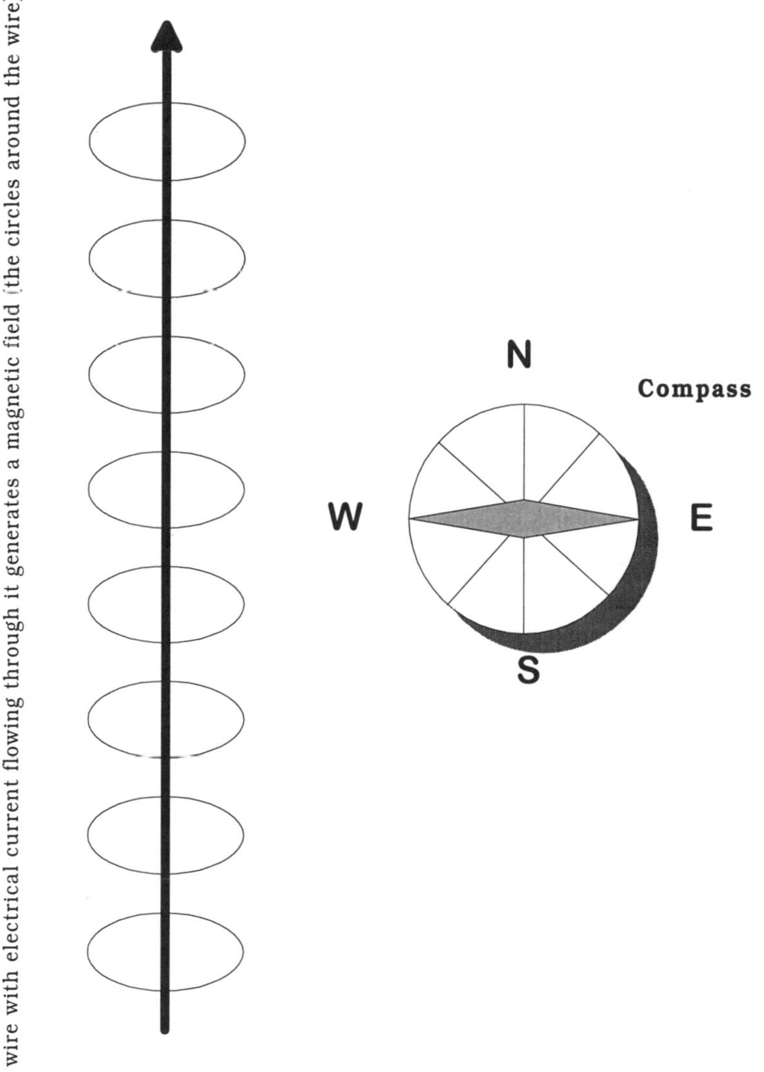

Figure 15. Hans Oersted discovered that a compass will detect the magnetic field generated by an electrical current flowing through a wire.

A burst of breakthroughs quickly followed. Soon after Oersted announced his discovery in 1820, French physicist Dominique-François Arago confirmed the magnetic effects of an electrical current passing through a wire. Then André-Marie Ampère demonstrated that two wires carrying electrical currents will either attract or repel one another, depending on whether the currents are running in the same or opposite directions. In 1821 the Russian-German physicist Thomas Seebeck discovered the phenomenon of thermoelectricity, the conversion of heat into electricity when the junction of certain metals is heated.

That same year, 1821, another English scientist published the first of several ground-breaking reports on magnetism and electricity. Michael Faraday reported on his discovery of electromagnetic rotation in a paper entitled "On Some New Electromagnetical Motions, and on the Theory of Magnetism." He used his discovery to build the first motors powered by electricity. Nine years later Faraday and American scientist Joseph Henry independently discovered the principle of the dynamo. Basically, Faraday and Henry both found that a moving magnet could create, or "induce," an electric current in a nearby wire. Just as electricity could create magnetism, magnetism could create electricity.

Henry's work had more immediate practical applications,[3] but Faraday's work had more far-reaching theoretical implications. For example, one of Faraday's discoveries about magnetism was based on an experiment that millions of children have done with a simple horseshoe or bar magnet and some iron filings. When placed in proximity to the magnet, the iron filings form patterns of lines in the space surrounding the magnet. Some invisible force emanates from the magnet and fills the space around it (Figure 16), forcing an alignment of the iron filings. If you pick up the magnet, you cannot feel anything surrounding it. Put it to your tongue, you taste nothing but metal. You can hear no sound coming from it, smell nothing different about it. You certainly do not see any "aura" or halo surrounding the magnet. It appears to all your senses to be a simple metal bar.

But the iron filings tell you differently. Invisible lines of magnetic force emanate from the magnet and fill the space around it.

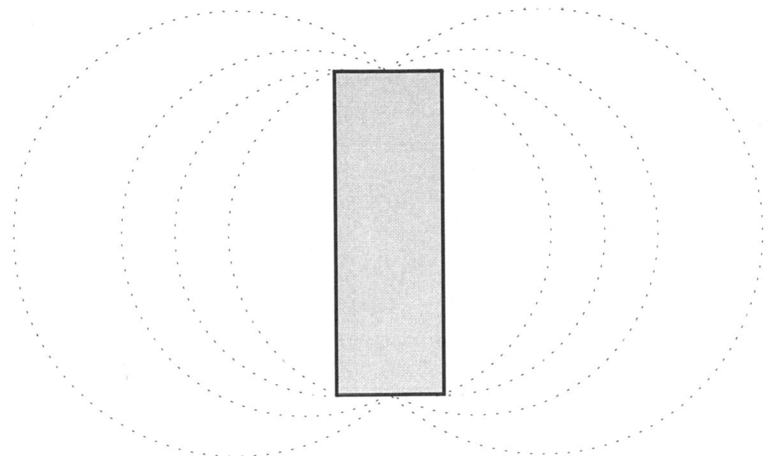

Figure 16. Michael Faraday discovered the existence of invisible "lines of force" surrounding a magnet.

Faraday mathematically assigned a number and a direction to every point in space surrounding the magnet. A freely suspended compass needle moving near a magnet will tend to point in a particular direction. The amount of magnetic force affecting the needle corresponds to a specific positive or negative number, depending on whether the pole that the needle is approaching is the "positive" or "negative" magnetic pole. Faraday was the first person to call this array of numbers and directions filling space a "field." The concept of force fields has played a major role in the development of modern physics.

In 1845 Faraday made another discovery that would have a profound effect on physics. In fact, it would eventually lead to the destruction of the Newtonian vision of reality. His discovery had to do with the polarization of light, the process of confining wave vibrations to a single direction. Suppose you and I are holding the two ends of a jump rope. You start flicking your end up and down, sending a series of waves down the rope. The waves are aligned vertically; they move up and down as they travel down the rope to my end. You stop the flicking and start snapping your end back and forth from left to right. That sends waves traveling horizontally down the rope. Suppose you then start flicking your end in all different directions, sending a flurry of waves down the rope. Some will be vertically aligned, some horizontally aligned, some at other angles. Now, suppose a third person comes along with two large pieces of wood. This person stands near the rope and places one piece to the left and one to the right of the rope, creating in effect a large vertically aligned slit. The rope's vibrations quickly become aligned vertically. This, essentially, is what polarization does.

Scientists had known for some time that certain crystals and other substances could polarize light. This was just another piece of evidence that light was made of waves and not particles. Faraday's discovery, however, was much more than a confirmation that light could be polarized. He found that *a magnetic field* could affect the way in which certain crystals polarize light. Faraday realized that there must be some kind of connection between light and electromagnetism. He went on to suggest that light might consist of waves in the electromagnetic lines of force.

William Wordsworth was still alive when Faraday made his discoveries about light and electromagnetism (Goethe, who had fought so hard against the Newtonian theory of light, had died in 1832). A beloved figure in England, Wordsworth was named poet laureate in 1843. He surely must have known about these breakthroughs. What might he have thought of these amazing discoveries about light? In his youth he had been a pantheist, seeing God as consisting of and existing in everything; his poems celebrated his love and worship of nature. For Wordsworth, the "I" and the "eye" were intimately connected, and the existence of invisible realities was an essential element of his Romanticism. Faraday's discovery of the "invisible reality" of magnetic lines of force, and of the connection between light and electromagnetism, would have appealed to a pantheistic, mystery-loving poet.

However, by the end of his life Wordsworth's theological outlook had become more mainstream. He was now a member of the Church of England and had long discarded his pantheistic proclivities. The poetry of his later years, for the most part, also fell far short of his earlier brilliance. It, too, had become "mainstream." By the time nineteenth-century science had taken the first steps toward the profound mystery of light and the quantum, almost all the Romantics were dead and Wordsworth had left Romanticism behind.

Other clues to a connection between light and electromagnetism began to pop up. For example, in 1857 the German physicist Gustaf Kirchhoff worked out the mathematical relationship between magnetism and electricity. He found that in doing so the speed of light was a constant in his formula.

The next advance came in 1864. James Clerk Maxwell followed up Faraday's ideas about the relationship between electromagnetism and light. Three years before Faraday's death Maxwell developed a set of mathematical formulas that described light as a series of electromagnetic waves. Maxwell published his formulas in a paper entitled "A Dynamical Theory of the Electromagnetic Field." His equations suggested that other forms of electromagnetic waves might also exist. His 1873 book *Electricity and Magnetism* laid out the basic laws of electromagnetism and also predicted the discovery of radio waves. Ten years after the publication of *Electricity and Magnetism,* the Irish physicist George Francis Fitzgerald pointed out that Maxwell's theory seemed to suggest that an oscillating electrical current could create a series of electromagnetic waves.

In 1888 Maxwell's theory of light as an electromagnetic wave received powerful experimental verification from Heinrich Hertz. Following Fitzgerald's suggestion, Hertz showed that an oscillating electrical current emitted invisible waves of electromagnetism: Hertz had discovered radio waves. Most people called them Hertzian waves until Italian electrical engineer Guglielmo Marconi, inventor of the radio, renamed them radiotelegraphy waves.

The Ether Crisis

Since light was a wave phenomenon, some transmission medium must permeate all of space. As noted at the beginning of our story, physicists referred to it as the ether. If the ether existed (as it must), then there must be some way to detect it. In 1875 Maxwell had suggested the development of a sensitive measuring tool that could detect the interference patterns in waves of light. Such a device is called an interferometer. Two American researchers took up the challenge. In 1888 Albert Michelson and Edward Morley built an extremely sensitive interferometer to measure how the earth moves through the ether. They made two simple and reasonable assumptions: (1) that the ether was stationary relative to the earth, and (2) that the earth moved through the ether as it circled the Sun. Their interferometer would measure the speed of the earth moving through the ether by measuring the velocity of light in different directions. As the earth moved toward the source of the light, plowing through the ether, the light's velocity would obviously be a bit faster than when measured in the opposite direction. The difference in speed, however tiny, would prove conclusively the existence of the ether and provide still more support for the wave theory of light. Michelson and Morley had every confidence of detecting the differences in light's velocity. They knew—as did every other physicist of the time—that light was a wave phenomenon. There could be no doubt about that. Their experiment would be the icing on the cake.

What Michelson and Morley found, however, was not what they expected, for their experiment, as extremely sensitive as it surely was, detected absolutely no differences whatsoever in the velocity of light. In the process the two scientists had come up with a way to measure the speed of light with unprecedented accuracy. But that was not what was important; they thought they had failed miserably.

Their results implied that Earth was standing still, but that was obviously incorrect. No one had believed that Earth was stationary since Copernicus's day. If light was a wave, it *must* travel through a medium like ether. But ether apparently did not exist. So how could light travel from the Sun to Earth? Could it be not a wave after all? How could *that* be the case when so much other observational evidence said it was?

The problem went even deeper than these obvious observational disparities. Since the end of the seventeenth century, scientists had believed that all physical phenomena could be explained by Newton's laws of motion. Newton's theory implied the truth of the principle of relativity, which states that the laws of the universe stay the same for all systems moving at constant speeds relative to each other. Newtonian physics, however, assumed the existence of absolute space. Young's discovery of the wave nature of light required the existence of the ether, which was a substance linked to the existence of absolute space. The ether was something that could be called a fixed frame of reference: something against which it was possible to determine the absolute motion of a moving body.

Physics found itself caught in a terrible conundrum because the Michelson–Morley experiment did not detect the ether. If the ether did not exist, perhaps absolute space did not exist, either. And how could that be? How could there not be *something* fixed in the universe, against which we can make our measurements and come up with firm results?

Because Hertz had discovered the existence of radio waves, another form of light that acted as a wave and was invisible to the eye, it was even more pressing that an explanation be found for the null result of the Michelson–Morley experiment. Two physicists offered one. George Fitzgerald and Dutch physicist Hendrick Antoon Lorentz presented a somewhat outlandish suggestion. If objects *contract as they move,* they said, the problem of light's invariant velocity disappears. The speed of light appears the same in different directions, they said, because Michelson and Morley's interferometer was contracting just enough in the direction of its motion to counteract the change in light's velocity. Meanwhile, the ether was still out there.

As it turned out, Fitzgerald and Lorentz's contraction idea was right, but not in the way they originally assumed. In 1905 Albert Einstein published two papers that together composed his special theory of relativity. In this theory Einstein stated that all the laws of physics must operate the same way in all frames of reference that are in steady motion with respect to one another. This in turn means that the speed of light must be constant, no matter how or where or how fast a body is moving. Michelson and

Morley got the right answer because there is no fixed frame of reference in the universe against which to measure the movement of Earth. They measured a constant speed of light because the speed of light *is* constant: 299,792,458 meters per second in a vacuum. The Fitzgerald–Lorentz contraction does take place, too, according to Einstein's theory. An object contracts along its length in the direction of its motion. The contraction does not become obvious until the object is moving at a significant fraction of the speed of light—but it happens, and physicists have detected it.

Einstein's theory meant that there was no need for any ether to carry light from one place to another. Light *always* moved at a constant velocity. Even though he probably knew of it, Einstein had never been very interested in the null results of the Michelson–Morley experiment. His concern had been with the validity of Maxwell's equations for electromagnetic waves in systems moving close to the speed of light. His special theory of relativity explained why Maxwell's equations were valid even in a frame of reference moving at close to the speed of light.

At the basis of special relativity were three assumptions:

1. Absolute speed cannot be measured—only speed relative to another object.
2. No matter how fast an observer or a source of light is moving, the speed of light will always be the same.
3. No material object in the universe can ever travel faster than the speed of light.

Ever since the formulation of the special theory of relativity, some scientists and a lot of science fiction writers have been trying to find a way around these three postulates. Warp speeds and superluminal jumps abound in science fiction movies and books. In those fictional realities, anything goes—as long as the writer can get the reader to suspend disbelief. In the real world, however, those three pillars of special relativity stand unshaken. Nearly a century of experiments and observations have failed to find any chinks in the armor of these assumptions. In particular, the speed of light is always constant, and nothing possessing mass has ever been clocked moving faster than the speed of light. The bumper sticker that announces "186,000 miles per second: It's not just a good idea, it's the law," is both humorous and true. The ether has vanished from physics like fog burning off under the noon-day sun.

Einstein's papers on special relativity were only two of five he published in 1905 that rocked the world of physics. Even before the two special relativity papers, Einstein had come up with a good reason why the ether did not need to exist. To do so he had relied on the work of another German physicist, one who was trying to patch still another gaping hole in the supposedly perfect fabric of nineteenth century physics. The man's name was Max Planck, and the hole was the appropriately named the ultraviolet catastrophe.

The ultraviolet catastrophe had to do with something so commonplace in the nineteenth century that few people thought about it. Railroads were becoming a major method of long-distance transport. A few horseless carriages were running about by then, but most short-range transportation was still by horse or by horse and buggy; the blacksmith still had job security. Today few people have seen a piece of iron being heated up, but in those days most people knew what happened when you heated iron: It glowed. Few really wondered why an iron bar glowed red when it was first heated. But physicists couldn't find the answer.

We've already seen how astronomers and astrophysicists have uncovered the information encoded in light and used it to study the properties of distant planets, stars, and galaxies. The information hidden within light eventually triggered some of the "earthquakes" that have reshaped our vision of the cosmological landscape. Newton's discovery of the spectrum made this possible. Other researchers, such as Fraunhofer and Kirchhoff, learned about the spectra of incandescent gases, of heated bodies, and of light passing through clouds of cool gas. And all of us have seen the most common form of the spectrum, the rainbow. Many of us have seen still another form of the spectrum, but have not recognized it for what it is.

Ever since humans started heating iron and other metals to forge them into swords or plowshares, we have seen this form of the spectrum. As a nonburning solid like iron is heated, it begins to glow with a dull red light. As we heat it up still more, its color begins to change from dark red to bright red, then to orange and to yellow, and finally to an eye-hurting white.

As the electromagnetic wave theory of light became accepted, physicists naturally tried to find an explanation for this well-known phenomenon. Obviously, at lower temperatures an object like a piece of iron in a smithy's forge emits light of fairly long wavelengths. That's why its color is red or orange. As the iron rod gets hotter it emits light with shorter and shorter wavelengths. When it is white-hot all the wavelengths of light are being emitted. Astronomers noted that the same phenomenon seemed to be true of stars. The hotter the star, the bluer was its spectrum.

The question was why? Why did objects like hot iron pokers and stars act this way? What was the theoretical explanation for it? When the theorists of the time tried to come up with an explanation for this using the wave theory of light, they stumbled. And badly.

The Atomic Vision

To understand how the ultraviolet catastrophe led to quantum mechanics and the overthrow of the Newtonian vision of reality, we need to take a short but important side trip into the world of the atom. The word *atom* comes from the Greek *atomos*, meaning "indivisible." The idea that the world is made of tiny particles or entities that cannot be further subdivided is an extremely old one in Western culture. It goes back to an ancient philosophical dispute. Heraclitus, a Greek philosopher who lived in the late sixth and early fifth century BCE, believed that the basic law of nature was change. All things change, he said; that is the essential nature of everything in the cosmos. A fourth-century BCE philosopher named Parmenides, however, disagreed with this position. Parmenides held that reality was unchangeable. Change, he asserted, was merely an illusion.

A third Greek philosopher offered a compromise position. Democritus, who also lived in the early fourth century BCE, said that change was a part of reality, but that reality overall was unchangeable. Change for him consisted of the motion of otherwise unchangeable basic constituents of reality. These entities he called atoms, and they composed all being. Everything else was a void. Democritus's ideas about the different forms and actions of atoms turned out to be quite modern in some respects. For example, he suggested that atoms came in different shapes, and that this helped explain the varying properties of different materials. When atoms stuck firmly together, he said, they made solid objects; when they hung loosely together they made liquids; when they bounced off one another they made gases. The larger the atoms, the denser the object they composed. No size limit existed for atoms, Democritus thought. Atoms the size of a world could exist somewhere in the cosmos. Today we know that atoms *do* come in different "shapes" and "sizes." We call them elements. (The word *element* comes from the word the ancient Greeks used for the four fundamental building blocks of reality in their worldview: earth, air, fire, and water.) We also know that the firmness or looseness with which atoms "stick together" does indeed determine whether the substance they compose is a solid, liquid, or gas. And we even know of "atoms" the size of worlds: We call them neutron stars, and they are quite real.

Democritus is called the father of atomic theory. However, it's probably more accurate to call him the grandfather. Atomism as a viable theory of reality found few followers after Democritus. Epicurus of Samos, who lived in the third century BCE, founded a school of philosophy based on the idea of atoms. However, the big guns of Greek philosophy, Aristotle and Plato, rejected Democritus's ideas.

The atomic theory resurfaced in 1649. That year the French natural scientist Pierre Gassendi published a commentary on Epicurus's philosophy, in which he asserted that matter is made of atoms. In 1666 Robert Boyle's book *The Origine of Formes and*

Qualities also proposed that the universe was made of atoms. Like his rival Isaac Newton, Boyle had become entranced with the idea of a mechanistic universe. The idea of atoms—indivisible tiny particles—fit that view of reality.

The real breakthrough for the atomic theory came in 1803, from English chemist John Dalton. Dalton noted that when chemicals combine with one another to form new chemicals, they do so only in discrete, integral proportions. This meant, he said, that each chemical must in turn ultimately consist of specific indivisible parts—atoms. Four years later the third edition of Thomas Thomson's book *System of Chemistry* presented Dalton's atomic theory. In 1808, the French scientist Joseph Louis Gay-Lussac announced that gases combine chemically in specific proportions of volumes as well as in definite proportions of weights or masses. So gases, too, must be essentially atomic in nature.

By the middle of the nineteenth century the atomic nature of reality was becoming widely accepted in chemistry. Leading the way was the Englishman Humphrey Davy, whose writings and experiments had an enormous influence on the scientists of his day. His electrochemical research led to his discovery of the elements barium, boron, calcium, magnesium, potassium, and sodium. Davy also proved that chlorine was an element and theorized that the affinity of some chemicals for others was an electrical phenomenon. Other researchers, meanwhile, were also busy discovering new elements and publishing books and pamphlets on chemistry.[4] The rise of the atomic theory of matter—essentially a particle explanation of reality—was taking place at the same time that the wave theory of light was displacing Newton's particle theory.

In the 1830s Faraday made a proposal that presaged a major change in the atomic theory. Atoms were supposedly elementary particles. As their very name suggested, they could not be divided into any smaller pieces. Faraday, however, suggested that his experiments with electricity could be explained by assuming the existence of tiny particles of electrical charge attached to atoms. Like other researchers, he had recognized that while most atoms were electrically neutral, some carried either a positive or a negative electrical charge. In 1833 Faraday, along with William Whewell, proposed the name "ion" for electrically charged atoms.[5] The later success of Maxwell's equations describing electromagnetism as a wave phenomenon, however, obscured Faraday's suggestion.

The shift back to Faraday's idea was made possible by German inventor Heinrich Geissler. In 1855 he developed a pump powerful enough to remove nearly all the air from a closed glass tube. He used it to produce the first good vacuum tubes. A vacuum tube (also called an electron tube or a thermionic valve) is a glass tube containing two electrodes, electrical conductors such as a wire or a metal grid that either emit or receive an electric current. A positively charged electrode is called an anode, while a negatively charged electrode is called a cathode. Geissler and other fabricators of vacuum tubes quickly discovered that when an electrical current passed through the electrodes, a mysterious new form of radiation appeared in the tube, which caused what little air was left in the tube to glow, or fluoresce, thus betraying its presence. In 1858 the German mathematician and physicist Julius Plücker found that a magnet placed near the vacuum tube would cause the radiation within it to bend. It was clear that the radiation had some association with electrical charge. In 1874 George J. Stoney called these hypothetical electrical particles carrying this charge "electrons," and offered an estimate of the basic electrical charge of such a particle.[6] In 1876 Eugen Goldstein found that the radiation in a vacuum tube started at the cathode and passed across the vacuum tube to the anode. He used the term "cathode rays" for the light emitted by the radiation in the vacuum tube. During the 1870s the best vacuum tubes were built by the English chemist and physicist William Crookes. Similar tubes came to be called Crookes tubes.

In 1881 the German physicist Hermann von Helmholz showed that the electrical charges in atoms were not continuously distributed but instead divided into specific integral portions. This strongly suggested that the electrical charges must be particlelike in nature. However, not everyone went along with this resurgence of interest in a particle explanation of electricity. In particular, in 1883 Hertz carried out a series of experiments that led him to believe that Goldstein's cathode rays were waves rather than electrically charged particles. Nine years later he showed that cathode rays can penetrate thin films of metal.

The controversy over the nature of cathode rays is a classic example of how scientists can use two different models to explain most, but not all, of the observable facts. On the one hand, a group of German researchers that included Hertz believed that cathode rays were a peculiar form of electromagnetic waves because the rays could pass through thin metal films, something no particle was known to do. On the other hand, a group of mainly British scientists were equally sure that cathode rays were beams of tiny electrically charged particles. A magnetic field would bend a cathode ray, something that could not be done to any electromagnetic wave. The crucial missing piece, it turned out, was Hertz's experiments of 1883. He had tried to use an electrical field to bend a beam of cathode rays and had been unsuccessful. He had therefore concluded that cathode rays were not particles but waves. In fact, Hertz was wrong. An electrical field *will* bend a beam of cathode rays, and J. J. Thomson did just that in 1894. Thomson's experiment proved that the velocity of cathode rays was much slower than that of light. If cathode rays were waves rather than particles, therefore, they were certainly not the same as electromagnetic waves, which traveled at the speed of light.

The resolution of the cathode ray controversy came in 1897, just three years after the deaths of Hertz and Helmholz. Thomson devised an experimental apparatus that used a combination of magnetic and electric fields to bend the beam of cathode rays passing through a Crookes tube. This made it possible for Thomson to prove that cathode rays were beams of tiny particles: electrons. He was also able to measure the electric charge of the electron and, because of the amount of deflection of the beam, the mass. Electrons turned out to have a mass some two thousand times less than that of a hydrogen atom, the simplest and lightest of all elements. Electrons were subatomic particles. Thomson and others concluded that the electrons came from the atoms of what little gas was left in the Crookes tube. Electrons, in other words, were negatively charged pieces of atoms.

A Desperate Man and a New Vision

A glowing bar of iron, heated to incandescence in a blacksmith's forge: It was a simple and commonplace image, a phenomenon from the everyday life of nineteenth-century Europe. Surely a simple scientific explanation existed for it. By this time, as the new century lay only a few years away, physicists felt a surge of confidence in their profession. The wave theory of light, proposed in its then-modern form by Young, was almost totally triumphant. Maxwell's equations were superb examples of physics at its best. The scientists of the time must have felt like Archimedes, who supposedly once said of the lever, "Give me a place to stand and I will move the earth!" Physics had an enormously powerful lever in the form of Maxwell's equations and the wave theory of light. They felt themselves standing on solid ground. They felt they could move the entire universe.

Still, there was that iron bar. The problem of the glowing iron bar—and of other heated objects—is today called *blackbody radiation*. As we have seen earlier, a true blackbody does not exist, but some substances come pretty close. Lampblack, for example, absorbs more than 98 percent of all the radiation that hits it. Also, an object does not have to be solid to be a blackbody. It can also be a cavity inside a solid object. For example, we can take a metal box that is completely sealed except for a small hole in one side. When we heat the box until it begins to glow red and then look inside the tiny hole, we will see red light emitted from inside the box. This, too, is a blackbody phenomenon. The thermal or heat radiation emitted by a blackbody at any particular temperature is called blackbody radiation.

To explain the phenomenon of blackbody radiation, and thus the glowing iron rod, scientists turned to the prevailing model of the atom. Thomson's discovery of the electron led him in 1898 to propose what might be called a "raisin pudding" model of the atom (Figure 17). According to this model, the atom consisted of a positively charged main body (pudding) in whose surface were embedded negatively charged electrons (raisins). Add a few electrons, or take a few away, and you had some of Faraday's ions.

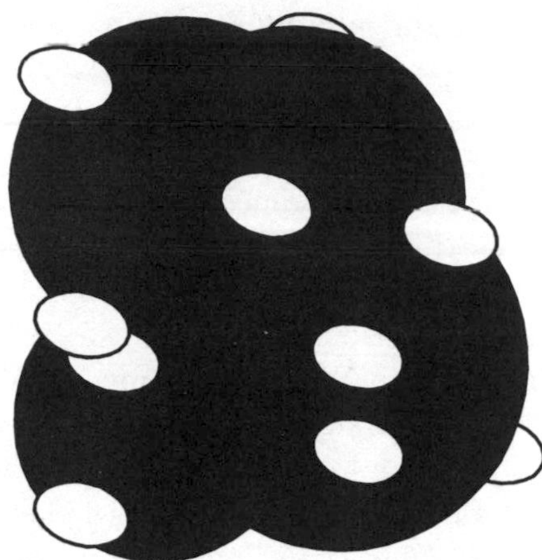

Figure 17. Thomson's "raisin pudding" model of the atom: a positively charged main body with negatively charged electrons embedded in the surface.

When an object was heated up, the heat energy caused the atoms to jiggle around faster and faster. That in turn made the electrons jiggle up and down. As they did so they would emit radiant energy in the form of visible light, in much the same way that Hertz's electrical oscillations made Hertzian waves. This in turn caused objects to glow as they heated up.

A simple explanation. The problem was, it didn't work. If this model were true, physicists soon realized, then heated atoms and their electrons would behave in ways that real-life observation said they did not behave. Even objects heated to only moderately high temperatures would emit immense amounts of blue, violet, and ultraviolet radiation. At very high temperatures they would emit nearly infinite amounts of radiation. But with their own eyes people could see that this was manifestly not the case. Objects heated a modest amount emit red light, not blue or violet light. Extremely hot objects emit only a finite amount of ultraviolet light. The name "ultraviolet catastrophe," therefore, was right on target.

Though physicists failed to find a way to make their classical theories explain blackbody radiation, it was not for lack of trying. The best-known attempts were made by Wilhelm Wien and John

William Strutt, Lord Rayleigh in the 1890s. Each tried to find a mathematical formula that could explain blackbody radiation, and each almost succeeded. Wien's formula worked at short wavelengths but not at long wavelengths. The formula derived by Rayleigh and Sir James Jeans, known as the Rayleigh–Jeans law, worked at long wavelengths but not at short.

In 1894 Max Planck also set out to discover a way out of this conundrum. And like his confreres he failed—at first. By 1897 he knew he was in trouble, and began casting about for a different approach to the problem. Using mathematical techniques first developed by Maxwell and Ludwig Boltzmann for the statistical study of the movement of gas molecules, Planck in early 1900 independently worked out Wien's formula. But that formula, as we've seen, turned out to be incorrect at explaining blackbody radiation at long wavelengths.

At this point Planck was getting a bit desperate. Perhaps out of desperation, or out of near-poetic inspiration, Planck made a radical new supposition about the nature of light. Like every other physicist on the planet at the end of the nineteenth century, Planck had been assuming that when atoms were heated up, their electrons emitted energy smoothly and continuously. As long as the electrons were "jiggling" they were emitting energy in this fashion. According to the classical Newtonian view of reality, all energy flows continuously, without breaks, like water flowing in a stream. To make his equations work, to make them accurately predict the emission of blackbody radiation as it occurred in the real world, Planck had to make an apparently absurd assumption: Suppose that energy does *not* flow continuously? Specifically, suppose that an object like a blackbody—or a near-blackbody, like a heated iron bar or box, or a particle of lampblack—emits radiation only in discrete amounts, in tiny spurts? In the same way, a blackbody or similar object would only *absorb* energy in discrete units.

Planck described the process by which such spurts of energy were released by a blackbody as "quantizing." The energy was released in tiny packages by "quantized oscillators" (the jiggling atoms). (The word *quantum* was not new; it had been used as far back as 1567 and meant "quantity" or "amount." However, this was the first time it had been used in physics.) Part of Planck's explanation of blackbody radiation was a new mathematical constant he invented to make the math work. Symbolized by *h*, Planck's constant[7] is used to calculate the size of the energy packet for each frequency of light (the number of light waves that pass a particular point each second). The color of a particular beam of visible light is directly related to the light's frequency. Planck described this relationship between energy and frequency with the mathematical formula $E = hf$, where E is energy, f is frequency, and h, Planck's constant, is the constant of proportionality between them.

Planck showed that the energy packets, or quanta, for any specific color of light are the same size. Moreover, the quanta for different colors differ in size. The quanta of low-energy light like red are smaller than those of yellow, green, or blue light. When a blackbody is heated up, it first emits quanta of light with low energies. As the heat increases, the object can emit quanta with greater and greater energies, the quanta that constitute blue and violet light. To our eyes, the change in color seems continuous. The reason, says Planck's mathematics, is simple. Quanta are inconceivably small, so small that our eyes cannot discern their discrete existence. We see only a continual change in the color of the iron bar as it heats up, from red to yellow and on smoothly to eye-hurting white. But in the land of light the quanta exist.

Planck announced his solution to the blackbody radiation problem in December 1900, the last month of the last year of the nineteenth century. It was appropriate, even though he didn't know it. An old vision of reality was coming to a close and a new one about to unfold. Light may be made of waves, as Thomas Young and others asserted. But Max Planck had provided a mathematical explanation that worked. Planck's equations solved the ultraviolet catastrophe, but the solution seemed absurd. To explain the smooth and continuous release of a steadily changing rainbow of light from a hot iron bar, Planck's formula required that light was emitted in a granular, discontinuous fashion. Not only was apparently smooth matter "granular," made of tiny, discrete objects called atoms, but so was light. So was all electromagnetic radiation. No matter what our eyes might say about light, or our fingers might say about flowing water, both matter and energy are made of pieces.

Reality, examined at very close range, was beginning to look very strange. The outline of a new alternate reality was taking shape.

Einstein Meets Planck

Another difficulty involving light remained—the nagging problem of the existence or nonexistence of the ether. The first step to its resolution had come in 1887, when Heinrich Hertz discovered that certain metals emitted electrons when a beam of light was shone on them, the photoelectric effect. The number of electrons emitted by the metal depended on the intensity of the light. Intensity, in physics, is the rate at which radiant energy strikes a unit of area; the kinetic energy—the energy of movement, or velocity—of the electrons is *not* connected to the light's intensity. That's a surprise! Common sense would dictate that the more intense the light beam, the greater the rate at which the light energy hits the metal, and the more kinetic energy the electrons would have. What is even more surprising, however, is what *does* affect the kinetic energy of the electrons: What counts is the *color* of the light. In other words, the shorter the wavelength, and thus the higher the frequency of the light hitting the metal, the greater the electrons' kinetic energy. That makes no sense at all. Light is a wave, for heaven's sake. Why should the distance between the wave peaks have any effect on how fast the electrons boil off the metal?

Planck's 1900 paper had simply focused on the blackbody problem, and his invention of the idea of the quantum was really meant to apply only to that issue. Neither Planck nor anyone else had seriously thought of applying the quantum concept to anything else in physics. Einstein, however, saw things a bit differently. He knew of Planck's paper and saw in it a way to explain the photoelectric effect. He did so by borrowing Planck's idea of the quantum. His explanation was based on the assumption that light sometimes acts as if it is made of particles and sometimes as if it is made of waves. One way to picture this is to imagine that, if they are examined with an extremely powerful microscope, the waves that compose light turn out to be "granular," as if made of particles. (This is a metaphor, by the way—an "as if." It is not what light "really" is.) Electrons are knocked loose from the metal plate only when they are directly hit by incoming light particles. The greater the light's intensity, the more photons hitting any unit of surface. Remember, the energy of the photon has only to do with the light's frequency, and not its intensity. The higher its frequency (and thus the shorter its wavelength), the more energetic are the particles that compose it. And how do we determine just how much energy each photon is carrying? Why, said Einstein, we'll just take that constant (h) from Planck and multiply the light's frequency with it. That reveals the amount of energy carried by the photon. Therefore, it is possible for increases in light's frequency to increase the kinetic energy of the electrons knocked from the metal plate by the light beam.

In applying Planck's constant to the problem of the photoelectric effect, Einstein did something just as revolutionary as Planck had done five years earlier. Planck had described the *process* by which blackbodies release and absorb energy as quantized. Einstein, though, used h to show that energy itself comes in quantum chunks. Planck had shown that the energy of light increases with its frequency, that is, that energy was proportional to frequency. Ultraviolet light, with its short wavelength and high frequency, carries a large amount of energy. Red light, with its long wavelength and low frequency, carries a smaller amount of energy. Einstein used h to show that the photon's energy is itself quantized. Light sometimes acts like waves and sometimes like particles. We find ourselves talking about particles of light using wave terms, and waves of light as if they were packaged into particles. It's totally absurd. And it is real.

Wheeler Arrives

The year 1911 was an auspicious year for physics in more ways than one. Not only was it the year that Wilhelm Wien won the Nobel prize in physics for his work on blackbody radiation (which in turn figured into Planck's invention of his eponymous constant), it was also the year John Wheeler was born in Jacksonville, Florida, the son of two librarians who strongly encouraged his interest in literature and science. By the time he was three years old he was asking his parents the tough questions like, "If I keep going out into space, will I ever come to an end?" As a youngster, he became fascinated with mechanical devices—and explosions. When he was ten he found a dynamite blasting cap and set it off. He lost the tip of a finger but not his interest in poking at the universe. After graduating from high school, he headed north to Baltimore and Johns Hopkins University. There he entered a fast-track program in physics and by age twenty-two had his Ph.D. He has been on the faculty at Princeton University since 1938, with numerous side trips into industry, government service, and appointments at other colleges and universities.

Approaching ninety, Wheeler is still active in body and mind as the Joseph Henry Professor Emeritus at Princeton University. His inquisitive imagination has peered deeply into the most intractable problems of modern physics. His visions are mind-boggling, and his superb communications skills have enabled him to present those visions to both physicists and the world at large. He invented the term "black hole" and was the first to postulate that space-time at the quantum level is like foam. Under his direction, one of his graduate students came up with the idea of multiple universes—not as science fiction, but as a valid alternate vision of reality. Wheeler is the main proponent of the "participatory universe"—the belief that our participation in the cosmos helps create reality. Unlike many of his colleagues, Wheeler does not shy away from the philosophical and metaphysical implications of quantum mechanics. Indeed, he boldly charges in where most other quantum physicists fear to tread.

As he moves through his eighties and the millennium nears its turning point, Wheeler is trying to apply information theory to quantum physics. The result is an exciting new glimpse at the very nature of reality. But all of Wheeler's work, including his speculations about quantum mechanics and information theory, rests on a foundation that started being built the year he was born. The man who laid that foundation was a Danish physicist whose brilliance of mind rivaled that of his friend and friendly competitor, Albert Einstein.

From Planck to Bohr

Planck's explanation of blackbody radiation created enormous waves (pun intended) in physics. He had provided a solution to the ultraviolet catastrophe, but it was a purely mathematical solution, one that no one could *visualize*. Planck himself was very uneasy about it. Knowing he had dropped a rather large rock in the pond of physics, he tried to back away from the implications of his work. But the formula was out there. Planck's constant had been loosed on the world. When Albert Einstein used h to explain the photoelectric effect, the quantum was more than a curious mathematical constant. It had some kind of reality to it. In 1913 a Danish physicist named Niels Bohr would make use of it again.

As noted earlier, J. J. Thomson's discovery of the electron had led him to propose the "raisin pudding" model of the atom, which held that the atom was like a positively charged pudding studded with negatively charged electron raisins. By 1911 Thomson—now Lord Thomson—was head of the famous Cavendish Laboratory in England and the main promoter of the raisin pudding atom. A hydrogen atom had one electron raisin embedded in pudding. Pull out the raisin, said Thomson, and you had a positively charged hydrogen ion. Helium ions could carry a double positive charge, so obviously the helium atom had two electron raisins that could be plucked out of the positively charged helium pudding. And so on.

However, Thomson's model was not the only one for the atom. The other major model had first been proposed by Philipp Lenard and Japanese physicist Hantaro Nagaoka. Lenard's experimental evidence showed that an atom was mostly . . . nothing; empty space—as far from the raisin pudding model as one could get. In 1903 he suggested that atoms are really pairs of electrons, and some still-unknown positively charged subatomic particle somehow bound them together in mostly empty space. The following year Nagaoka offered his own model, postulating that atoms consisted of a positive core with electrons circling it in a ring, like the planet Saturn.

Which model was correct? In 1907 Ernest Rutherford, like Thomson a well-known British physicist, set out to answer the question. Along with his associate Ernest Marsdan, Rutherford had set up a series of experiments to test the theory of the raisin pudding atom. To do so was fairly easy. One had only to bombard materials with beams of charged particles (like electrons, for example) and see how the materials deflected the particles. Rutherford was at that time the leading expert in the then-new field of radioactivity. Back in 1898 Marie and Pierre Curie had discovered that the element thorium spontaneously emits what they first called "uranium rays." Marie Curie soon renamed the phenomenon radioactivity. Rutherford later discovered that some materials gave off another kind of radioactive particle that he called alpha rays or alpha particles. Today we know that alpha rays are actually the nuclei of helium atoms.

Rutherford and Marsdan carried out a long series of experiments, bombarding a series of foil targets with alpha particles and observing how the particles scattered. They did so by watching the flashes of light produced by the scattered alpha particles as they hit a screen coated with zinc sulfide. If Thomson's raisin pudding model was correct, the positively charged alpha particles would be barely deflected by the diffuse positively charged pudding of the atom. The electrons would also have little effect on the alpha particle beam, since they were small and dispersed widely through the positively charged raisin pudding. The prediction, therefore, was pretty straightforward. If the raisin pudding model was correct, most of the alpha particles would pass right through the foil targets. The rest would be barely deflected and would come off the targets at very shallow angles.

What Rutherford and Marsdan actually observed was something they did not expect at all. Most of the alpha particles did indeed pass through the foil targets, and most of the rest of the particles were indeed deflected at only very shallow angles. However, a small but significant portion of the alpha particles were deflected at very *large* angles. In fact, some of them ricocheted off the foil targets and came right back at Rutherford and Marsdan. Rutherford later remarked that it was like firing an artillery shell at a piece of tissue paper and having the shell bounce off the paper and come flying right back at you. It was, indeed, a rather disconcerting experience.

Thomson's raisin pudding model could not be correct. Instead, the atom must consist of a very tiny positively charged core or nucleus with electrons occupying the rest of the space. Lenard's model didn't match Rutherford's experimental evidence, either; Nagaoka's came closer, but he had postulated thousands of electrons per atom, whereas the evidence was that atoms had only a few electrons each. Rutherford's model was more like that of a solar system than a planet with rings. The electrons appeared to orbit the central nucleus like the planets orbit the Sun.

However, this model had serious problems. It did not match the experiment. First of all, the negatively charged electrons and positively charged nucleus would attract each other. Somehow the electrons must overcome this attraction to stay in orbit. Moreover, electrons give off energy as they move; that had been proven, and it was the reason why iron bars glowed when heated. If electrons really did orbit the nucleus in this fashion, then they would be moving continuously along in their tiny orbits and continuously emitting radiation: All matter would constantly glow as it emitted electromagnetic radiation throughout the spectrum. However, the evidence of our own senses tells us that this is not true. Matter is made of atoms, and it's obvious that the objects around us are not continuously glowing with heat and light. Moreover, as the electrons in atoms emitted radiation they would lose energy and quickly fall into the positively charged nucleus. Most atoms are actually very stable; only a few are radioactive and change into other elements.

What, then, were the electrons doing in the atom? How did they stay a part of the atom without falling into the nucleus or continually radiating away energy? Niels Bohr came up with a solution to the problem by turning, as Einstein had done eight years earlier, to Planck's quantum.

The son of a physiology teacher, Bohr had received his Ph.D. in physics in 1911 in Copenhagen. He was offered an opportunity to work with Thomson at Cavendish Laboratory, and he eagerly took it. Bohr had written about Thomson's model of the electron for his thesis and had found some mathematical errors in Thomson's work. When he arrived at the Cavendish one of his first moves was to tell Thomson of his mistakes—not exactly the diplomatic thing to do. By the autumn of 1911 Bohr was on his way to Rutherford's laboratory in Manchester. Thomson was not sorry to see him go.

In 1913 Bohr started working on the serious problem facing Rutherford's solar system model of the atom. Like many other physicists at Manchester, Bohr wanted to find out where the electrons were in the atom, how they stayed there, and why they radiated energy as they did in the form of electromagnetic radiation. Most of the time they did not release radiation. What's more, if electrons were crashing into the atomic nuclei at the center, then atoms would be inconceivably smaller than physicists knew they were. Every object in the universe would quickly collapse. The earth would shrink to the size of a football stadium. Light would not exist, for there would be no moving electrons to release electromagnetic waves.

One thorn in Bohr's side was the nature of the spectrum of heated hydrogen gas. When a pure gas like hydrogen is heated up, it does not emit a rainbow spectrum of light from red through violet. The spectrum of a heated pure gas consists of a series of bright lines, called an emission spectrum. The emission spectrum of hydrogen had been discovered in 1880 by Swiss mathematician and physicist Johann Balmer. The lines of the hydrogen spectrum are today called Balmer lines, or the Balmer series. When Bohr in early 1913 first read Balmer's paper on the spectrum of hydrogen gas, he realized he might be able to find a way to explain the Balmer lines.

The problem facing Bohr was serious, but in its essence simple. He needed to find a model of the atom that allowed the electron to radiate energy in an on-again, off-again fashion rather than continuously. And the model had to provide a reason why electrons acted this way. The answer Bohr found lay in Planck's quantum. Bohr knew that the size of a quantum could be calculated. He began working out the quantum of hydrogen. With a positively charged nucleus and just one negatively charged electron, hydrogen is the simplest of all elements.

Bohr knew that both Planck and Einstein had used the Planck constant to show that matter—a heated bar of iron or a piece of photoelectrically sensitive metal—absorbs and emits light energy in piecemeal fashion. What Bohr did was use the Planck constant, h, to show that the very constituents of matter itself also acted in a discontinuous fashion.

Bohr thought of h as an electron's unit of angular momentum. Those of us who have ridden on a merry-go-round have experienced the sense of being "thrown outward" by the circling merry-go-round. That is the force of angular momentum.

An electron has angular momentum because it is moving around the nucleus of an atom. The electrical attraction of the negative electron to the positive nucleus is the force that tethers the electron to the nucleus. Bohr calculated the circular orbit for an electron with one unit of angular momentum. That unit was an integer (a whole number like 1 or 2) multiplied by h divided by 2π written as $\hbar$. His result matched with observation: He had calculated the size of a hydrogen atom with one electron in its lowest orbit. Next he calculated the orbit of an electron with two units of $\hbar$. It, too, was correct, as was his calculation for a hydrogen atom's electron with $3\hbar$. Bohr found that only electrons with whole units of angular momentum can occupy orbits in an atom. These whole units of angular momentum are the quantum numbers of the electron orbits.

With these calculations, Bohr had shown that the electrons in an atom occupy orbits around the positively charged nucleus, orbits that have only specific energy levels. The orbits, Bohr orbits, are quantized. As long as the electron is in a particular orbit, it emits no energy. But if an electron absorbs sufficient incoming energy, then and only then will it "jump" to the next higher orbit or energy level. Under certain conditions the electron also releases energy, in pieces, or quanta, and then "jumps" back down to an orbit of lower energy. The jumps are discrete and instantaneous. We call them "quantum jumps."

One has only to think a bit about the implications of Bohr's 1913 model of the atom to realize that he had developed a profoundly disturbing view of reality. The evidence of our eyes and other senses tells us that reality is a seamless fabric; matter is essentially smooth in nature; movement is continuous. True, light may sometimes act like a beam of particles, but it also behaves like continuous waves. Even time moves continuously from past through present to future. Newton's classical mechanical vision of reality is based on our commonsense experiences. Even his "action-at-a-distance" force called gravity acts in a continuous fashion. Planets move smoothly and constantly around the Sun; the Moon orbits Earth continuously; there is nothing jerky or stop-and-go about the way an apple falls to the ground.

Now Bohr was upsetting the Newtonian mechanistic applecart in a way no one had before. His mathematical model of the electron in an atom was not just good; it was *very* good. He could use it successfully to predict the sizes of different atoms. And that was what made his model so disturbing. Electrons were emitting electromagnetic radiation in packets, or quanta, of energy. But they were not doing so merely by their movement around the atomic nucleus. Nor did they oscillate up and down, as Thomson and others had envisioned in the raisin pudding model. Bohr himself had tried mathematically to model a continuous movement of electrons from one orbit to another, but the result had failed utterly to match the real-world emission of light by atoms. No, the electron released or absorbed energy by "jumping" instantaneously from one energy level to the next. Still, physicists knew from observation and experiment that a stream of moving electrons radiated energy. That's what the discoveries of Oersted, Faraday, and other brilliant nineteenth-century explorers of electricity and magnetism had been all about. So how could electrons inside an atom *not* radiate energy as they moved about an atomic orbit?

Bohr found the answer in the size of an electron's quantum jump. Sometimes the distance between atomic orbits was relatively small; other times the jump was a relatively large one. When an electron made a quantum jump that involved a relatively small change in orbital diameter, the results were very close to what classical Newtonian calculations predicted. Bohr called this the Principle of Correspondence. Quantum mechanics, he said, must give the same results as classical physics when applied to large systems. Only when the quantum jumps were relatively large did the results differ widely from those of classical physics. The mathematics of quantum mechanics worked just fine, and provided extremely accurate predictions about electrons, atoms, and the energy they released and absorbed. Those predictions agreed with classical mechanics when necessary, and didn't when not necessary.

The Newtonian vision of reality was valid—to a point. Bohr's quantum vision of reality was also valid—to a much finer point. If Bohr was right about quantum-jumping electrons, then reality was essentially discontinuous, granular, made of tiny bits of something very, very small. The world according to Bohr is completely different from what our senses say it is. Our knowledge of it as smooth and seamless is in fact an illusion caused by the scale of our perception; our senses cannot perceive its fundamental granularity. Bohr's mathematical wizardry had succeeded in brushing aside the illusory veils that hide the quantum nature of matter and energy from our eyes. The quantum reality remains hidden from our physical senses, but the inner eye of the imagination can manipulate this alternate reality the way a poet molds old words into new worlds.

Bohr's 1913 vision of the atom had its flaws. Physicists discovered that the lines in the spectra of gases like hydrogen had a finer structure that Bohr's atomic model using the quantum did not explain. In 1916 the German physicist Arnold Sommerfeld showed that the existence of a second quantum number explained the fine structure. Sommerfeld's second quantum number was based on the idea that electrons moved in elliptical rather than circular orbits around the atom's nucleus. That same year Robert Millikan confirmed the reality of Planck's constant as well as Einstein's theoretical explanation of the photoelectric effect using h.[8] A magnetic effect on the spectrum of light, the Zeeman effect, was later explained by the existence of a third quantum number. Finally, in 1925 it was discovered that electrons have a property called spin.[9] A fourth quantum number accounts for the effects of spin. These four quantum numbers are all that is needed to provide a precise mathematical description of an electron in its orbit in an atom.

In 1919, six years after Bohr had quantized the electrons in atoms, his mentor Ernest Rutherford finally identified the proton as the positively charged portion of the atom. The proton has a positive electrical charge with exactly the opposite strength of an electron, just as Thomson and others had predicted years earlier. However, the proton does not have the same mass as an electron. It is about 1,836 times heavier, in fact. Thirteen years later, in 1932, another British physicist, James Chadwick, would discover the third major subatomic particle, the neutron. A neutron has a mass slightly greater than that of a proton, but has no electrical charge. Together the protons and neutrons compose an atom's nucleus. Chadwick had begun studying radioactivity with Ernest Rutherford, but then moved on to the Cavendish, Thomson's laboratory. He was assistant director of the Cavendish from 1923 to 1935 and it was there he discovered the neutron.

Matter as Waves

Four years after the discovery of the proton a member of French royalty published a set of mathematical equations that sent new shock waves rumbling through physics. Louis Victor de Broglie was an honest-to-goodness prince who had originally planned on being a historian. His brother, however, himself a physicist, persuaded Louis to study physics. Louis soon became fascinated by the work of Einstein and Planck and with the wave–particle duality of light. De Broglie wondered if the opposite might also be true: that objects thought of as particles might also be described as having properties of waves. He was specifically thinking about the electron. De Broglie's Ph.D. thesis at the Sorbonne presented a mathematical formula that said that electrons and other subatomic particles can behave sometimes like particles and at other times like waves. Like light itself, matter has a dualistic nature, both wavelike and particlelike. De Broglie's thesis professor was quite dubious about this astonishing proposition. Before rejecting the thesis, however, he sent a copy to Einstein. Einstein's response was enthusiastically positive; de Broglie got his Ph.D. and a well-deserved place in the history of science. His equations set the stage for the next step in the development of quantum mechanics.

Now, of course, we are dealing with a picture of reality that is pretty hard to picture. Matter as a wave? It seems ridiculous. De Broglie and others imagined that the electron-as-wave was sort of like the waves caused when a tiny pebble is dropped in a circular bowl of water. The waves ripple out from the impact point, hit the circular rim and are reflected back into the center, where they bounce off one another and move back out to the rim . . . and so on. These kinds of waves are called standing or stationary waves. A standing wave is a form of wave whose profile does not move through its medium of transmission, but rather stays in one place. A traveling or progressive wave, by contrast, has a profile that moves through the medium at the speed of the wave. Sound waves, for example, are progressive waves. Large waves on the open ocean, on the other hand, are often standing waves. So are the waves in a violin string plucked by a violinist; the waves do not travel up and down the string but actually stay in one place on the string. They are standing waves.

De Broglie thus replaced the electron "particles" in Bohr's atom with electron waves. Alternatively, we can imagine the electrons as particles constantly accompanied by "shadows" that are waves. That's what made Einstein so enthusiastic about de Broglie's thesis. Einstein may have used Planck's quantum for his own purposes, but his purposes were classical in nature. He believed in the essential continuity of nature. De Broglie's wavelike electron had that kind of continuity to it.

The Next Quantum Leap

John Wheeler was only a very precocious fourteen-year-old and still in high school when quantum mechanics took the next leap. In 1925, the year that physicists discovered the electron's property of spin, Wolfgang Pauli proposed his exclusion principle. The exclusion principle states that no two fermions can exist in the same state at the same time. (A fermion is a class of subatomic particle that includes electrons, protons, and neutrons.) One of the consequences of the Pauli exclusion principle is that no two electrons in an atom can have identical sets of quantum numbers. Each electron's location in an atom is described by the set of four quantum numbers, and each electron in an atom must have a unique combination of those numbers. The result? The Pauli exclusion principle accounts for the distribution of electrons in all atoms. And it is the number and distribution of the electrons in an atom that determines each element's unique chemical characteristics. Chemical bonds between atoms are electrical bonds between the atoms' electrons. Pauli thus showed that an aspect of quantum mechanics governs the single most pervasive physical characteristic of reality. For it is the chemical characteristics of elements and of the molecules they combine to form that create the sensory tapestry of the cosmos. Smell, taste, texture, and sight: These senses all detect attributes of objects that are determined by their chemical properties.

What would Wordsworth and the other Romantics have thought of this? Intense sensory experiences gave rise to powerful emotions that, remembered later and alchemized in the poetic imagination, turned into extraordinary poems: verbal and written visions of reality. And now the despised scientists had uncovered the profound quantum mystery that is the foundation of all sensory experience. All of our senses use the electrochemical activity of nerve fibers and the brain to gather information about the outer world. Our senses of taste and smell are actually chemical senses; they depend on direct chemical interactions between molecules and neurons. All chemical processes are the interactions of the electrons of atoms. And those interactions are governed by quantum mechanics. To the degree that our individual personalities are shaped and changed by our experiences, then, our very self, our "I", is a quantum phenomenon. This is the kind of scientific knowledge that might well have excited a poet like Wordsworth.

At about the same time as Pauli was formulating his exclusion principle, a German physicist younger than the idea of the quantum burst on the scene. Werner Heisenberg discovered that he could use arrays or patterns of quantum numbers to determine the lines in a spectrum. He had begun walking the path into quantum mechanics in 1922, when he first met and talked with Bohr at a conference in Gottingen, Germany. By this time Bohr had left Rutherford's lab in England and was head of a new institute of physics in Denmark that his colleagues informally called the "Copenhagen School." In his book *Physics and Beyond*, Heisenberg recalls taking a walk with Bohr after one of the latter's lectures. "My real scientific career only began that afternoon," says Heisenberg, as he realized that "atoms are not things."[10] Heisenberg was one of many young physicists who were eager to make a complete break with the old Newtonian worldview. His encounter with Bohr led Heisenberg to question whether anything in the subatomic realm of reality could be truly *known*, truly *seen* for what it is. Heisenberg's mathematical construct was called wave mechanics. His vision of the quantum landscape used a method for organizing data into tables or arrays that had first been developed in 1859 by an Irish mathematician named W. R. Hamilton. This method was called a matrix, and Heisenberg's model has since come to be called matrix mechanics.

The same year that Heisenberg developed matrix mechanics and Pauli his exclusion principle, an American physicist named Clinton Davisson uncovered dramatic experimental proof for the existence of de Broglie's matter-waves. Davisson worked for Bell Laboratories, a corporate lab that has given the world a raft of practical and theoretical breakthroughs in science and technology. Davisson was doing some experiments involving large crystals of nickel and beams of electrons. He discovered that the electrons in the beam were reflecting off the surfaces of the crystals in patterns he could not explain. He published his results. Two German physicists saw the paper and its accompanying interference patterns and realized the patterns could only be caused by the matter-waves of electrons bouncing off the nickel atoms in the crystals. Davisson had thus confirmed de Broglie's theory of the electron's wave–particle duality.

The following year was as much a watershed in the evolution of quantum mechanics as 1925 had been. An Austrian physicist named Erwin Schrödinger was utterly fascinated by de Broglie's idea that electrons could be thought of as waves as well as particles. In January 1926 he used de Broglie's concept of standing waves to develop a mathematical equation that explained every shift that occurred in the electron wave patterns within an atom. Schrödinger developed his own set of wave equations to determine the location of spectral lines in light; Bohr had developed his mathematics empirically to explain the experimentally observed placement of hydrogen's Balmer spectral lines.

Schrödinger faced the same question that Michelson and Morley had encountered in trying to detect the presence of light waves in the ether: Just exactly what is "waving"? Schrödinger wasn't sure, but he gave it a name anyway. He called his mathematical wave functions by the Greek letter *psi* (pronounced "sigh"), or Ψ.

When Schrödinger's wave equation was applied to the hydrogen atom, it gave the precise energy levels for the electron that Bohr's model had predicted. Moreover, the wave equation also applied to more complicated atoms and even to electrons not bound within the atom. As long as a subatomic particle was not traveling near the speed of light (relativistically, as Einstein would put it), Schrödinger's wave equation produced exactly the right description of the particle's behavior.

Schrödinger's wave equation basically says that matter can and does act like a standing wave. An atom, for example, can propagate as time passes through different patterns of standing waves. Schrödinger's wave equation allows physicists to calculate the order through which these patterns of standing matter-waves propagate. If a physicist knows the initial state of a particular atom, he or she can use the Schrödinger wave equation to predict precisely the pattern of its propagation. And given the same initial state, the atom always follows the same sequence of standing wave pattern propagation. Schrödinger soon realized that his wave equation and Heisenberg's matrix mechanics were simply two different mathematical ways of saying the same thing. Heisenberg was saying "po-TAH-to" while Schrödinger was saying "po-TAY-to." No one called anything off. It was far too late for that; the quantum cat was out of the bag.

Schrödinger believed that his mathematical equations described real entities. An electron is, well, an electron. It's not quite like anything we see in our daily lives—a mountain ash tree, a kangaroo, a billiard ball—but it is nonetheless real. Schrödinger imagined that an electron was a kind of foggy, tenuous critter. It was smeared out over its wave pattern like a splotch of paint on a canvas. This image is quite simplistic. A hydrogen atom, with its one electron, has a standing wave with three dimensions (length, width, and height). But a helium atom's two electrons are standing waves existing in 6 dimensions, lithium's three electrons are in a total of 9 dimensions, and a uranium atom's ninety-two electrons exist in 276 dimensions! Most of us can handle three dimensions. Far fewer of us can deal with Einstein's famous general relativity concept of a four-dimensional space-time continuum, in which time is as real a dimension as the three spatial ones with which we are familiar. But 276 dimensions? No one can envision what *that* looks like.

Three other developments in quantum mechanics also occurred in 1926. A few months after Schrödinger developed his matter-wave equations, Paul Dirac mathematically derived Planck's law from first principles, a powerful achievement that thrust the twenty-four-year-old English physicist to the forefront of his field. Dirac would go on to put the finishing touches on quantum mechanics as we know it today. In doing so he would also reveal the existence of antimatter. Then, in October 1926, an American chemist named Gilbert Lewis introduced the word "photon" as the name for the quantum of light.

The final advance that year in quantum mechanics came from Max Born. Born addressed a problem in Schrödinger's wave equation. One way to imagine the wavy electron of Schrödinger's theory is to return to the image of the jump rope. Suppose one end of the rope is tied to a wall. If I suddenly flick my end of the rope I send a wave pulse traveling down the rope to the other end. The traveling wave will hit the wall and bounce back up the rope. Now if you, standing next to me, throw a tennis ball at the wall, it, too, bounces back. My wave pulse acts like your tennis ball, and vice versa. Schrödinger's equation described the electron as a kind of wave and showed the different patterns that wave would take as it expanded in time. As long as an electron stayed inside an atom, the electrical forces of the nucleus tightly confined its wave patterns. The waves that "shadow" the electron can only spread out exactly as far as the atom is wide.

If the electron is free, however, and not confined inside an atom, the picture becomes quite different. Now nothing acts to confine the electron wave. The mathematics of Schrödinger's equation says that it will expand, beating out its sequence of wave patterns through time and space. Within a millionth of a second the electron will be the size of a football stadium. The problem with this picture, of course, is that it has no correspondence to reality. When we do detect electrons, they are always like tiny dots. Indeed, physicists today consider electrons to be as close to one-dimensional entities as one can get. They have no internal structure and can be treated as mathematical points in space that move through the one dimension of time.

Born suggested a way out of this conundrum, but the cost was enormous. He proposed that one way of interpreting Schrödinger's equations was to say that they gave the *probability* of an electron being located in a particular orbit with a particular momentum and position. In other words, either Schrödinger's and de Broglie's waves were "real things" or they were abstract mathematical concepts "into which we cannot enter."[11] According to Born, the electron or any other subatomic particle is not real, not in the sense that we perceive a bowling ball, or a ball bearing, or even a fleck of dust as real. Rather, the electron or any other subatomic particle is a *probability wave*.

Born's interpretation sounded the death knell for classical physics. Not only was matter wavelike as well as particlelike, but one of the basic constituents of all of "reality," the electron, was not real. It was instead a set of probabilities. The classic vision of reality as constituted of real objects in concrete relationships to one another had dissolved into a vision of probability ghosts and shadows. But more was still to come from the minds of the quantum mechanics.

A Vision of Uncertainty

At the time Schrödinger and Heisenberg were developing their specific versions of quantum mechanics, no one had seen an atom: Powerful electron microscopes did not yet exist. And in any case, an atom and its constituent parts were as much waves as particles. Heisenberg's matrix mechanics and Schrödinger's wave equation said so. Moreover, Heisenberg was now asserting that all we really know is the state of an atom or electron at the beginning of an experiment and its state when the experiment is completed. We can never know what's "real" in between these two states. The land of the quantum, said Heisenberg, is forever closed to our inspection, and we should forsake any attempts to create models of what goes on there.[12]

In 1927 Heisenberg took the next philosophical and scientific step forward from his vision of quantum reality. Armed with his matrix mechanics and his conviction (based on his math) that the quantum landscape can never be seen or fully mapped, Heisenberg proposed what has come to be known as the uncertainty principle. As he originally proposed it, Heisenberg's uncertainty principle had to do with the electron. It is impossible, said Heisenberg, to determine both the position and the momentum of an electron at the same moment in time. (Momentum in physics is the measure of a body's motion that is equal to its mass times its velocity.) One quantity of this pair might be known with great accuracy—its position or its momentum at any particular moment—but only at the cost of great uncertainty in knowing the measurement of the other quantity.

Suppose you want to measure these two properties of an electron, its position and its momentum, with as much accuracy as possible. You want to know where it is *and* the product of its mass and velocity. So you shine a light on the electron. What kind of light, though? You will need light with a very short wavelength and a corresponding high frequency to see the electron. But the photons of light with high frequency carry a great amount of energy and momentum. When a photon of high-frequency light hits the electron it will change the electron's momentum in a way you cannot possibly control. You may know with great precision the electron's position, therefore, but you cannot know its momentum with any precision at all.

Heisenberg had discovered that there are limits to what we can know about quantum entities like electrons. By "know" he meant "measure accurately." The farther we venture into the land of the quantum, the less certain become our measurements of certain characteristics of quantum entities. Our vision starts getting blurry, as it were. This "blurriness" or uncertainty about measurement is not caused by the limited visual capacity of our eyes or the weakness of our other sensory abilities. Nor is it caused by any imperfections in any measuring instrument we use as an extension of our senses. Nor is it due to the infinitesimal size of the entities we are trying to measure. What Heisenberg showed with this mathematics, rather, was that this "blurriness" is an essential, inescapable aspect of the quantum landscape itself. Reality itself, at the level of quantum entities and quantum events, is *inherently uncertain*.

This doesn't make any sense at all to us who live in the "real world." Give me the right kind of ruler and a stopwatch and I can tell you, with great precision, the position and speed of that car moving down Alabama Street. I touch Isis, my cat, and I know she is real. She is right here—well, no, she just jumped up and rushed over to the corner, where she is now staring at something unseen on the wall. But she is nonetheless real, and sitting over there.

Suppose the uncertainty principle operated in our everyday world of the senses, and it applied to two characteristics of our normal visual ability: color and shape. If it did, then the world we see with our eyes would be a lot different than it is today. Simply put, we could clearly see either the shape of objects or their color, but not both at the same time. You can squint *this* way and the colors of that mountain ash blossom forth in brilliance. Red berries, tan leaves (it's winter now), dark brown bark, the white of the snow piling up on both: beautiful. But you cannot make out the shape of the tree at all. It is nothing but a blur of color. However, if you look at the tree in this other fashion, its shape suddenly leaps out at you with crystalline clarity. You can see every leaf, and every crinkle on every leaf. You see the rough pattern of the bark, the smooth roundness of each and every berry. You can even discern the granularity of the teeny snow piles atop the berry clusters. But what you see is all in black and white and shades of gray. It's one or the other; you can never see both at the same time.

The Alternate Reality of Schrödinger's Cat

Erwin Schrödinger provided the world with a brilliant thought-experiment that illustrates the bizarre consequences of quantum mechanics. His thought-experiment is called Schrödinger's cat. It reveals that the weirdness of quantum mechanics applies not only to the land of the quantum, but also to the landscape we inhabit on a daily basis. The two realms are inextricably entangled with one another.

Like a jazz riff, Schrödinger's original formulation of his cat conundrum has been modified and added to over the years. Here's one of the better-known versions:

Suppose we take a cat and put it in a box. The box has plenty of air, so the cat will not suffocate. We also put in the box a special device that contains one radioactive atom, an electric circuit, and a vial of deadly poison. Then we close the box.

The radioactive atom is vitally important to the paradox of Schrödinger's cat. Many atoms have different versions, or isotopes. The isotopes of an atom all have the same atomic number—the same number of protons in their nuclei—but have differing atomic weights and mass numbers because they have different numbers of neutrons in their nuclei. Some isotopes are stable; others are unstable or radioactive. Marie and Pierre Curie had discovered that some atoms decay into other atoms by emitting various kinds of subatomic particles or fragments of atoms, a process called radioactivity. Each kind of radioactive atom has a specific half-life. The half-life of a radioactive material is the amount of time it takes for half of the atoms in the sample to decay into their final nonradioactive element. The carbon atom has three isotopes, of which two are stable and one is radioactive. Uranium, however, has six isotopes; they are all radioactive, and each has a different half-life. Some half-lives are measured in billions of years. For example, uranium-238 (which physicists and chemists abbreviate as ${}_{92}U^{238}$; the 92 is the atom's atomic number and the 238 is its atomic weight) has a half-life of 4.5 billion years. Given any specific amount of uranium-238, in other words, after 4.5 billion years one half of it will have decayed into a final nonradioactive atom (in this case, an isotope of lead). Uranium-235, which is used in atomic bombs, has a half-life of 700 million years. Carbon-14, which is used for dating many very old organic objects, has a half-life of 5,730 years. Magnesium-56 has a half-life of about two and a half hours. Gallium-68, an isotope of the element gallium, has a half-life of 68.3 minutes. Some isotopes even have half-lives measured in mere fractions of a second.

For the purpose of our thought-experiment with the cat, we are using a single atom with a half-life of one hour. This means that in sixty minutes there is a 50 percent chance that the atom has emitted a subatomic particle, in this case a gamma ray. We arrange the mechanism so that when the atom decays, the gamma ray it emits will cause an electric circuit to close. That in turn will cause a tiny pin to pierce the vial, releasing the deadly gas and killing the cat.

And here is the crux of Schrödinger's thought-experiment. We wait one hour. What does the box then contain: a dead cat, or a living cat? If we open the box we'll know; but without looking in the box, do we know what has happened to the cat?

A Newtonian classical physicist will tell you that the cat is either alive or dead. It is in one state or the other, and when we open the box we will know what that state is. And that state *is*, whether we open the box or not. But quantum physics says this is not the case. What *is* the case, however, is open to interpretation. As we shall see, how we interpret the puzzle of Schrödinger's cat opens a Pandora's box of new alternate realities.

Consciousness and the Measurement Problem

The Schrödinger's cat paradox is an example of what quantum physicists have called "the measurement problem." The equations of quantum mechanics make it possible to calculate the wave functions for photons, electrons, atoms, and other denizens of quantum reality. We can determine different possibilities of existence for such quantum entities. And as we saw, Schrödinger and Heisenberg revealed that such an entity, like an electron, will go on unfolding its different wave function patterns as time progresses.

When we make a measurement of the electron, however, something "magical" happens. All the possibilities for the electron's momentum or position collapse—all but one. That's the one that is measured. The collapse of the wave function occurs, as far as we know, instantaneously, at the moment of observation. So the question now is, When exactly does the wave function collapse from possibility to actuality? Or to put it another way: Who does the observing? We've come now to the famous "tree in the forest" question that comics love to play around with, that philosophers make a living with, and that physicists try their best to avoid thinking about: "If a tree falls in the forest and there's no one present, does it make any sound?"

A couple of possible answers exist to this question:

- If someone, anyone, is in the forest near the tree, it makes a sound when it falls. The wave function collapses when someone hears.
- The tree makes a sound when it falls even if there's only a tape recorder nearby. The wave function collapses even if the measurement is made with an inanimate instrument.

One way of examining the measurement problem is to go back to the experiments done more than a century ago by Thomas Young. As noted earlier, Young was the physicist who convincingly showed that light was a wave phenomenon. His experiments involved shining a beam of light through two closely spaced slits in an opaque barrier onto a luminescent screen. By doing this Young observed a set of interference patterns that is characteristic of waves.

Now, suppose we do a similar experiment using a special light source that emits one photon at a time. The spacing of the slits and the aim of the photon "gun" are such that any photon emitted will pass through either one or the other slit. Photon detectors will signal if the photon passes through slit A or slit B.

The photon source emits a photon. The photon, remember, is a quantum entity. It consists of "something" that sometimes acts like a particle and sometimes like a wave. Because it is a quantum entity, as Einstein proved, we can use the mathematical equations developed by Heisenberg, Schrödinger, and their colleagues to calculate its possible states of being before a measurement occurs. The photon's wave function contains two patterns. One is the possibility that it will pass through slit A, and the other is the possibility that it will pass through slit B. What happens next?

- You enter the room and look at the measuring device.
- A few moments later, I enter the room and ask you what happened.

- Later that day my wife asks me about the results of the photon experiment.
- Fourteen months later my paper on the results of the experiment appears in a prestigious physics journal, read by hundreds of scientists around the world.
- Two years after that a world famous physicist-turned-science writer publishes a best-selling book about quantum physics that is read by hundreds of thousands of people. One of them is a famous movie director who decides to make a documentary explaining quantum physics to the masses.
- Five years after our experiment the film wins an Academy Award for best feature-length documentary. Millions of people have seen it and learned about our experiment. One of them is a young physicist who after seeing the film goes home, goes to bed, and has a dream about inventing a faster-than-light drive for spaceships using some aspect of our experiment that had never occurred to anyone before. She wakes up at 4 a.m. and scribbles down the equations that danced in her dream.
- One hundred years later our grandchildren are colonizing the third planet circling Alpha Centauri.
- A half-million years later the human species has colonized thousands of planets across the Galaxy and is now leaping across space to the Large Magellanic Cloud.

What happened here?

Newton, Einstein, and other classical physicists would say that the photon gun emitted a photon. The photon had an equal chance of passing through each slit. It traveled across space, passed through one of the slits, and hit one of the detectors. If someone had wished, they could have followed the path taken by the photon.

Bohr, Heisenberg, and the other creators of quantum physics would reject this depiction of what happened. According to their creation, quantum mechanics, all that "existed" was a wave function, which in turn is not real but just a depiction of possible realities for the photon. The photon *in utero* could pass through slit

A or slit B. It had passed through neither until an observation was made, a measurement taken.

But, when did the "measurement" occur?

- When the detector detected the photon?
- When you checked the measurement device to see which detector had been triggered?
- When I asked you for the result?
- When my wife asked me for the result?
- When the journal editor read my paper?
- When the famous writer learned of the result?
- When the movie director found out which detector had fired off?
- When the young physicist saw the movie?
- When she had the dream?
- When my great-grandson, standing in his cornfield on Alpha Centauri 3, talked about it with his wife?

More than fifty years after Schrödinger first proposed his cat conundrum, John Wheeler summed up the bizarre implications of the measurement problem in a series of provocative questions: "Does observation demand an irreversible act of amplification such as takes place in a grain of photographic emulsion or in the electron avalanche of a Geiger counter? Does the quantum theory of observation apply in any meaningful way to the 'whole universe'? . . . How are observations made by different observers to be fitted into a single consistent picture of spacetime?"[13]

One thing Wheeler is firm about is the role of consciousness in quantum phenomena. " 'Consciousness,' " he has said, "has nothing whatsoever to do with the quantum process."[14] All that is required is what he, Bohr, and other quantum physicists call "an irreversible act of amplification." Some kind of "registration" of the quantum phenomenon is required, some record of the collapse of the wave function, a record that is indelible.

Consciousness is not necessarily indelible. We really do not *know* if consciousness survives death. And less dramatically, our memories are remarkably plastic. It is not difficult, using persistence and patience, to get a normal adult to "remember" something that never happened to him, or to forget events that did happen. The "irreversible act of amplification," then, might be the darkening of a grain of silver compound on a piece of photographic film, or a trace on a piece of graph paper, or electronic records impressed on a computer's hard disk. That is what Wheeler calls "measurement." Consciousness adds *meaning* to the measurement. But according to Wheeler, meaning, as we understand it, has nothing to do with quantum phenomena.

In the end, the entire universe is connected in some fashion to our experiment with the photon and the two slits. The measurement problem ultimately becomes one hell of a philosophical question: *Who is the observer of the universe*? If the wave function is not "real" until a measurement collapses its myriad possibilities into one observed actuality, *who collapses the wave function of the universe into actuality*?

Alternate Reality: The Copenhagen Vision

In 1927 Born's probability explanation of Schrödinger's wave mechanics and Heisenberg's uncertainty gave birth to the most commonly accepted philosophical interpretation of quantum mechanics. That year the world's leading physicists met in Brussels at the Fifth Solvay Congress on Physics. One of the questions they considered was, essentially, how to philosophically interpret the message of the quantum equations. It was now clear that Schrödinger's wave mechanics and Heisenberg's matrix mechanics were different ways of saying the same thing. It was also clear that the equations of quantum mechanics were incredibly accurate at both statistically describing the overall behavior of quantum entities and predicting the probable behavior of individual quantum entities. The question, though, was what did it mean? What, exactly, was being described by quantum mechanics?

The result of this debate has come to be called the Copenhagen Interpretation, mainly because it was the philosophical position championed by Niels Bohr, a Dane. Einstein, the last of the great classical physicists, opposed Bohr's interpretation. He well understood the implications of quantum mechanics and could not accept them. "God," he once said, "does not play dice." Nearly everyone else at the Congress, however, had come to recognize that God *does* play dice, and constantly. It was simply no longer possible for them to deny the power of the quantum equations, or the accuracy of their description of . . . whatever. Einstein was simply unable to let go of old visions and embrace new ones.

Bohr had no such problem. He plunged into a realm where few physicists had dared tread since the rise of rationalistic science in the eighteenth century. Bohr asserted that the world we sense is certainly real. The reality of phenomena and entities on a large scale—of birds and bees, people, rocks, hurricanes, oceans, forest fires, planets, stars, galaxies, and "great walls" of galactic superclusters—is revealed to us by a narrow band of electromagnetic radiation we call visible light. It is also the world we smell, touch, hear, and taste. It is the world of nature, a nature not only "red in tooth and claw,"[15] but also one with a "sense sublime / Of something far more deeply interfused, / Whose dwelling is the light of setting suns. . . ."[16] But Bohr went on to assert that the landscape of reality revealed to us by our senses, as real as it is, does not rest on a solid foundation. In fact, there *is* no foundation, said Bohr. There is no reality beneath reality. The realm of the quantum is not real. *It does not exist.*

The reason there is no quantum reality, said Bohr, is because quantum phenomena do not exist apart from observation. We cannot know what the quantum world is like because, in a sense, it does not exist until we carry out an observation. Only then is whatever we have observed "real" in the way we commonly mean.

Fifty-two years later John Wheeler would quote his old mentor Bohr to summarize this position in his typically pithy fashion: "No elementary phenomenon is a phenomenon until it is an *observed* phenomenon."[17] He would later write that "what we call reality consists of a few iron posts of observation between which we fill in an elaborate papier-mache of imagination and theory."[18]

More playfully, Wheeler has also invoked the old joke about three baseball umpires discussing how they call balls and strikes.

The first umpire says, "I calls 'em like I sees 'em." The second umpire says, "I calls 'em the way they *are*." And the third umpire says, "They ain't *nothin'* till I calls 'em!"[19] The first umpire is most of us, living in our everyday world. The second umpire is Albert Einstein, who insisted that an objective reality existed within quantum reality and that we need only find the "hidden variables" in quantum mechanics to make it come out like his beloved classical physics.

The third umpire is Niels Bohr—and John Wheeler. "They ain't *nothin'* till I calls 'em!"

Quantum theory, Bohr insisted until his death in 1962, is *not* an "alternate reality"; it is nothing more or less than a set of mathematical tools for predicting the behavior of atoms and subatomic particles and describing what we see. And it works. That's all that counts. The actual nature of the entities we are measuring is not important. The wave function of Schrödinger is not reality, either. It is merely the description of reality that the quantum physicist uses. To ask what the wave function *is*, said Bohr and his followers, is meaningless.

Science has always been based on the assumption that "reality" exists "out there," beyond the bony box of our skulls and the brain tissue within. The whole point of science has been to develop theories or explanations of the phenomena we see, theories that have some approximation to "the truth" that is "out there." Some explanations are better than others. The better explanations are those that come closer to "the truth." More accurate theories replace and subsume less accurate theories. So Newton's theory of gravitation overturned Aristotle's, and Einstein's general theory of relativity incorporated Newton's gravitational theory within it. Rutherford's model of the atom replaced Thomson's, and Bohr's replaced Rutherford's. We might not ever be able to experience the truth of reality directly—to gaze straight at the face of God, as Moses dared—but at least we could see it through a mirror, obscurely. At least we knew there was an objective reality "out there."

Then Schrödinger and Heisenberg had the effrontery to try to explain electrons and atoms. And the whole picture dissolved. For the Copenhagen Interpretation of quantum mechanics says there is no real correspondence between theory and reality, at least not on the quantum level. For quantum reality does not exist until we measure something.

Quantum Reality and Idealism

Observer-created reality is a characteristic of a philosophy known as idealism. This philosophy basically states that all entities and phenomena are representations of the mind. These representations, these creatures of mind, may well reside in some higher plane of existence than we perceive in our daily lives. Plato was the first and greatest philosopher to develop the concept of idealism. He held that certain universal objects or concepts were the only true realities. The entities and concepts we perceive in our daily lives are just imperfect copies of those universals—shadows cast on the back wall of a cave. The universals existing outside the cave in the realm of ideals are what's real. In the fourteenth century William of Occam also propounded an idealistic philosophy. George Berkeley, an eighteenth-century Irish-English Anglican bishop and theologian, developed a more modern version of idealism. He suggested that the individual consciousness is the source of ideals and that nothing exists independent of perception. "*Esse est percipii*," Berkeley wrote. "To be is to be perceived."[20]

However, Berkeley was a churchman, and a believer, and he held that it is the eternally existing mind of God that makes all of reality real. Moreover, Berkeley was essentially talking about what Wheeler has called "multiple quantum processes." The simple act of picking up a cup of coffee, for example, involves untold numbers of quantum phenomena. Hundreds of quadrillions of atoms in our fingers almost touch an equal number of atoms in the handle of the coffee cup. Many trillions of photons enter our eyes and stimulate rods and cones in the retina, sending millions of electrochemical nerve impulses into our visual cortex. Niels Bohr was talking about the *individual* quantum process: individual photons traveling through Thomas Young's slits in an opaque sheet of paper and hitting a fluorescing screen, say, or a single gamma ray emitted by a radioactive isotope in a box with a cat.

In the eighteenth century, Emmanuel Kant reconciled rationalism and empiricism by pointing out that reasoning and experience go hand in hand. He held that we may gain knowledge of the entities and phenomena of our physical world but that other entities lying beyond our experience were unknowable. This form of idealism is called transcendental idealism. The existence of "noumena," or "things-in-themselves"—truth and beauty, for example, or freedom, or God—that exist in this transcendental reality can be neither confirmed nor denied, nor can science ever prove they exist. The nature and existence of such entities are the purview of the branch of philosophy called metaphysics. In his book *Critique of Pure Reason*, Kant declared that scientific thought cannot provide answers to the questions surrounding these noumena.

In the early nineteenth century, Goethe and the German philosopher Georg Wilhelm Friedrich Hegel championed a form of idealism that saw all of reality as a creation of spirit or mind. Their positions were largely a strong romantic reaction to the rising tide of scientific materialism. Hegel believed that using the mind to intuit the existence of a priori concepts was superior to experimentation, while Goethe tried to uncover scientific explanations by using general philosophical principles. More recent proponents of the philosophy of idealism include the English philosopher Francis H. Bradley, whose lifetime spanned the end of classical physics and the rise of quantum mechanics, and the Italian philosopher and historian Benedetto Croce. Croce's 1902 book *Aesthetic as Science of Expression and General Linguistic* is a landmark of the modern formulation of idealism.

However, it would be a mistake to say that the Copenhagen Interpretation is a form of idealism, or that it confirms some essential truth of idealism. Every version of idealism begins with the assertion that ideals or universals somehow "exist out there." These universals have some sort of real existence, even though science cannot confirm or deny it. The "out there" is some other realm of reality, or the human mind (which may be "in there" rather than "out there"), or the mind of God. But the Copenhagen Interpretation denies even these presuppositions. *There is no landscape of the quantum*, says Bohr. If you cannot measure "it," then "it" is simply unimportant. It doesn't matter if ideals or universals exist at all. What's important is making the observation. What is observed is real. "No elementary phenomenon is a phenomenon until it is an *observed* phenomenon," says Wheeler. Everything else is imagination and theory.

The Copenhagen Interpretation, in fact, comes closer to pragmatism. This method of philosophy states that the truth of a proposition is measured only by its correspondence with experimental results and by its practical outcomes. Charles S. Peirce and William James are the founders of the philosophy of pragmatism. Virtually unknown during his lifetime, Peirce believed that only by examining the consequences of an idea can we know the idea's meaning. His major essays appeared in the posthumous collection *Chance, Love and Logic*, published in 1923. William James is the well-known psychologist and philosopher whose brother was the equally important novelist Henry James. Like Peirce, William James held that the truth of a proposition is knowable only by the proposition's consequences. He rejected all transcendental principles—and at the same time followed a lifelong fascination with religion and psychic phenomena.

Another way to try to grasp the implications of the Copenhagen Interpretation is to consider what it has to say about Schrödinger's cat. We have put the cat in a closed box with a mechanism that, after one hour, has a fifty-fifty chance of killing the cat. The question is, then, Is the cat alive or dead? The Copenhagen Interpretation of quantum physics is that Schrödinger's cat is in a kind of limbo. Until we open the box and look inside, it is neither alive nor dead, but a bit of both. The cat is represented by a quantum wave function, a probability wave of extreme complexity (after all, this is a cat and not an electron) that grows and expands through its various patterns until some entity carries out an observation. That quantum-cat-wave function carries the possibility that the cat is alive and the possibility that the cat is dead. When we make the observation one of those possibilities suddenly, instantaneously, becomes actuality, and the other possibility vanishes into nothingness: The wave function for the possibility that did not occur collapses.

This interpretation of the state of Schrödinger's cat is utterly bizarre from our commonsense vision of reality. We know that the cat is in the box, and we know from our daily experience that it is either alive or dead. Classical physics tells us in no uncertain terms that we learn about objects by observing them and that the objects we observe are real. But that is not what quantum physics says. Quantum physics tells us, in the equally uncertain terms of its enormously powerful and successful equations, that something is not "real" *until* we observe it.

Alternate Reality: Many Worlds

John Wheeler has been in the thick of path-breaking physics since his mid-twenties, when he began earning an international reputation as a first-class theoretical physicist. One of his early physics papers, for example, was entitled "The Mechanism of Nuclear Fission"—surely one of the most understandable and self-explanatory titles ever given a theoretical physics paper. It presented the definitive theory for the mechanism of nuclear fission. Wheeler played an important role in the Manhattan Project, America's successful effort to develop a nuclear fission bomb. Later he worked with Edward Teller on the development of the H-bomb.

Wheeler has also mentored many young scientists who later went on to make significant contributions to physics. One was Richard Feynman. As we saw earlier, Feynman developed a version of quantum mechanics called quantum electrodynamics, or QED. QED is a theory that can calculate the behavior of electrons and other subatomic particles more precisely than even Schrödinger's and Dirac's work.

Feynman was not the only one of Wheeler's graduate students to make a significant impact on quantum mechanics. Another was Hugh Everett III. In the late 1950s Everett produced a doctoral thesis that proposed a simple but radical way out of the Copenhagen Interpretation's measurement problem. The result was a new philosophical interpretation of the nature of quantum reality.

Everett's theory solves the measurement problem quite simply. The Copenhagen Interpretation says that only one of the possibilities in a wave function actually becomes real, and that it occurs only when a "measurement" or "observation" takes place. All the other possibilities contained in the wave function collapse. According to Everett, however, *every* possibility inherent in any wave function is real, *and all of them occur*! Each possible result of each wave equation takes place. The possibilities become actualities with each observation or measurement, multiplying at nearly infinite speed and branching out like some incomprehensibly complex tree of universes. With each observation or measurement, each and every possibility within the quantum wave function becomes real. That in turn means that slightly different realities, true alternate realities, come into existence with each choice, each observation, each measurement of a quantum event. It's not surprising that Everett's interpretation of the quantum world is called the "Many Worlds Interpretation."

Let's go back to Schrödinger's cat. An hour has elapsed. According to the Copenhagen Interpretation, the cat is neither alive nor dead, but in a "strange kind of physical reality"[21] that lies between actuality and possibility. When you open the box and look in, one of those two possibilities collapses into nothingness. Only *now* is the cat alive or dead. According to Everett's Many Worlds Interpretation, however, when you open the box neither possibility disappears. The quantum wave function of the cat bifurcates; the cat is alive and the cat is dead. You may see that the cat is alive, but at that instant another reality also exists. It is an alternate reality in which you see that the cat is dead. Both realities are real; both exist; both continue on their ways.

Consider again the experiment with the photon and the two slits. The photon can go through slit A or slit B. Now, suppose that if you observe that the photon has gone through slit A (the detector for slit A has registered the photon's impact) you send a letter of resignation to the head of the laboratory. If the photon goes through slit B, you take your letter and burn it. You cannot do both; you either send the letter or you don't send it. The Copenhagen Interpretation sees these two possibilities as mutually exclusive. However, the Many Worlds Interpretation says that when the photon's wave function collapses, reality splits in two. In one reality you send the letter and resign. In the other, you burn the letter. Each is real, but mutually exclusive. The "you" of alternate reality A, where the photon passed through slit A and you sent the letter to the lab's director, cannot encounter the "you" of alternate reality B, where the photon passed through slit B and you burned the letter. And this is just a simple example with two possible choices. Most of our daily choices are of the multiple type, with many different possible consequences. Suddenly, reality is very complicated indeed.

The Many Worlds Interpretation also provides a solution to another famous paradox of quantum reality called "Wigner's Friend." This particular paradox was first proposed by the Nobel prize-winning physicist Eugene Wigner. Wigner was fascinated by the role that consciousness supposedly plays in quantum mechanics and the collapse of the wave function in quantum reality. He was one of several eminent physicists who ascribed to the "consciousness creates reality" version of the Copenhagen Interpretation. This version says that an "observer" must be conscious or sentient. Reality is "the harvest of the quiet eye," to quote Wordsworth from his poem "A Poet's Epitaph." Or as he wrote in "The Excursion," it is created by the "eye . . . busy in the distance, shaping things . . .," though now the distance is not outward but inward, into the heart of quantum reality.

Other scientists who have embraced this position include Hungarian physicist and mathematician John von Neumann and American physicist Henry Stapp. When the tree falls in the forest, say these physicists cum philosophers, it makes a sound only if a conscious observer is present.[22] Wigner's quantum scenario was meant to show the central role that a *conscious* observer, such as a human being, plays in quantum mechanics. The paradox of Wigner's Friend goes something like this:

Wigner has invited his friend to come over to his house and perform the Schrödinger's cat experiment. The friend accepts the offer and in a room at Wigner's house he proceeds. The cat's in the box with the radioactive atom and the vial of poison. The cat is described by its quantum wave function, a standing wave pattern inside the box. As long as it's in the box and the friend has not opened the lid, the cat exists only in the strange "possible" land of the quantum. Its wave function describes it as possibly dead and possibly alive. Then Wigner's friend opens the top of the box and looks in. The cat is alive; its wave function has collapsed.

However, the friend has a surprise in store. Wigner opens the door to the room and walks in. He tells his friend that he, Wigner, has also been carrying out an experiment—on his friend! For that mirror over on the wall is really a one-way mirror, and Wigner was the observer. As long as Wigner had not looked through the one-way mirror into the room, his friend's observation of the cat existed as two different possibilities. In one the friend saw that the cat was alive. In the other the friend saw that the cat was dead. Only when *Wigner* looked into the room did the friend and his observation that the cat was alive become "real." In other words, Wigner's friend was *also* in a box, and Wigner collapsed his wave function. The friend and the cat owe their very existence to Wigner's conscious observation. And if that is the case, just how many boxes are there, and who is observing whom?

Wigner felt that the answer to this paradox lay with the first observer and his or her conscious observation of the cat's state—or the position or momentum of an electron, or whatever. Wigner's friend, said Wigner, created the cat's state of existence when he opened the box and looked in it. So Wigner, looking into the larger box of the room through the one-way window, observed an event that was already created by his friend.

However, Wigner's solution to his own paradox cannot be proven by the mathematics of quantum mechanics. So the problem of the "nested Chinese boxes" remains. Everett's Many

Worlds Interpretation resolves it in simple and sweeping fashion. All the boxes are real, and all possibilities exist in a universe with an infinite number of branches and leaves.

The Many Worlds Interpretation neatly sidesteps several disturbing problems associated with the Copenhagen Interpretation of quantum reality. One such problem is that of consciousness or sentience. Some quantum physicists, such as Wigner, believe that tape recorders and inanimate photon detectors aren't enough to collapse the wave function into the landscape of the everyday world of beaches, mountains, baseball games, and earthquakes. Only humans or other sentient beings can do it. In the Schrödinger's cat problem the quantum wave function of the cat collapses into a dead or living cat only when you or I or some other human opens the box and looks inside. A robot or android[23] will not do. In the photon experiment, does the photon's quantum wave function collapse when the detector detects the photon, or when a conscious observer (you or me) examines the detector? Wheeler and Bohr say the former; Wigner and his supporters say the latter. The Many Worlds Interpretation, however, says that consciousness is pretty much beside the point. It doesn't matter, because all possibilities inherent in the quantum wave function become real when *any* observation is made by anyone or anything—robots and tape recorders included.

Another related difficulty that the Many Worlds Interpretation deftly resolves is the measurement problem. As we saw, the measurement problem is summed up in the question: Who is the observer of the universe? That is, who collapses the wave function of the universe into actuality? Everett's take on this is no one; the universe has an infinite number of mutually inaccessible branches, and all of them are real.

The Many Worlds Interpretation is a fascinating one, and it is one that, mathematically at least, is perfectly consistent with equations of quantum mechanics. However, it is not a provable theory. No way exists to perceive these alternate realities or universes that come into being with each observation of a quantum event. Until some way is found to actually test or observe these alternate realities coming into existence, Everett's Many Worlds interpretation of quantum mechanics will remain on the borderland of science and imagination.

Wheeler's Visions

For all of his work in defense-related physics problems, John Wheeler's most far-reaching work has been in theoretical physics. Since the conclusion of his work on the H-bomb, Wheeler has spent nearly all his working (and probably waking) hours in a dynamic dance with relativity, gravitation, and quantum physics. It is in these rarified realms that Wheeler's greatest gift shines through: his ability to see reality in ways that no one else has before and to offer his visions to the rest of us with metaphors and images so striking that we, too, enter into his alternate visions of reality. He can capsulize in a few cogent sentences or phrases a wealth of information and inference about relativity, astronomy, and the remarkable implications of quantum physics for the nature of reality. In 1968, for example, Wheeler coined the term black hole when he wrote: "The light [from the surface of a superdense collapsing star] is shifted to the red. It becomes dimmer millisecond by millisecond, and in less than a second too dark to see. . . . [The star] like the Cheshire cat fades from view. One leaves behind only its grin, the other, only its gravitational attraction. Gravitational attraction, yes; light, no. No more than light do any particles emerge. Moreover, light and particles incident from outside . . . go down the black hole only to add to its mass and increase its gravitational attraction."[24]

Wheeler's most entrancing and provocative visions have dealt with quantum reality. In his continuing attempts to explain the kind of thinking he does about quantum physics, Wheeler has turned to one of the simplest yet most elegant experiments that shows the true weirdness of the quantum realm, the beam splitter experiment (see Figure 18). It begins with a beam of photons (or, if you prefer, an electromagnetic wave) entering a specially arranged array of mirrors and photon detectors. The beam first strikes a half-silvered (H-S) mirror set at a 45-degree angle to the

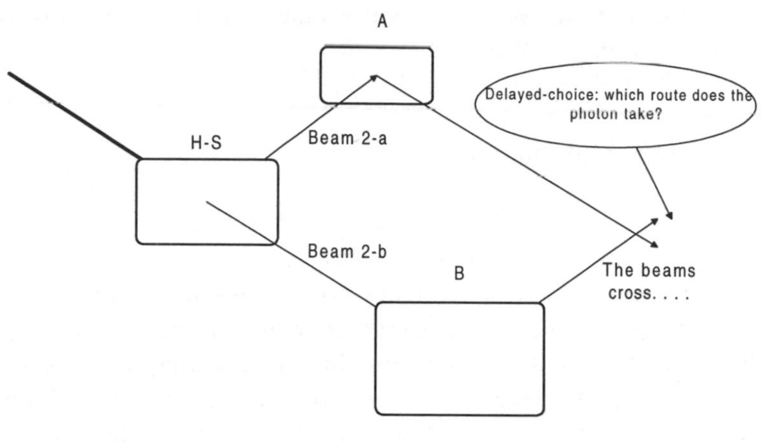

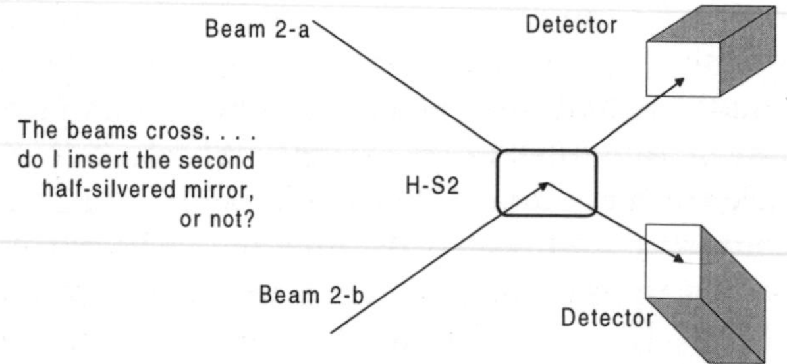

Figure 18. John Wheeler's delayed-choice thought-experiment: What happens when mirror H-S2 is inserted or removed at the very last instant?

beam. Such a mirror has the property of reflecting half the photons striking it and letting the rest pass through it as through clear glass. The beam is now split into two halves. The first half of the beam, which we'll call beam 2a, reflects off the H-S mirror and then strikes a regular mirror (called A) lying parallel to the H-S mirror. The rest of the beam, now called beam 2b, passes through the H-S mirror and hits regular mirror B, also lying parallel to the H-S mirror. The result is that beams 2a and 2b reflect off mirrors A and B and cross at 90-degree angles to one another at a final point.

Beam 2a enters one photon detector and beam 2b enters a second photon detector. Both detectors will signal the arrival of photons travelling along paths 2a and 2b, each beam with half the intensity of the original beam.

Now, though, suppose we place a second half-silvered mirror (call it H-S2) exactly at the point where the two beams 2a and 2b cross before hitting the detectors. Now what happens? What happens is that only one detector registers the arrival of the beam! It seems that one side of the second half-silvered mirror brings the beams into destructive interference. Remember: light can act as a wave as well as a beam of particles, and waves can interfere with one another and cancel each other out. The result in this case is that on one side of mirror H-S2 beams 2a and 2b cancel out and the detector registers nothing. The other side of mirror H-S2, however, brings beams 2a and 2b into constructive interference: The light waves reinforce each other. The result is beam 3, the fully reconstituted original beam of light, which hits the second detector. Only the second detector registers the arrival of the light beam, and at its original full intensity.

The business of only one detector registering the beam of photons in this latter case may seem odd at first, but it's perfectly reasonable. Even in light of classical physics it makes sense, for classical physics can adequately explain the destructive and constructive interference of waves. The action gets really strange, however, when we start looking at *one photon at a time*.

If we remove mirror H-S2, the second half-silvered mirror, from the apparatus, we'll see one detector or the other registers the arrival of a photon, as each successive photon arrives by either path 2a or 2b. But suppose we put mirror H-S2 back into the apparatus. Now only the second of the two detectors will fire each time a photon arrives at the end of the apparatus. The only way this can take place is if each photon *travels both path 2a and 2b simultaneously*. Destructive wave interference keeps the first detector from ever registering a photon, while constructive interference makes sure the photon always registers in the second detector.

This was the kind of thought-experiment[25] that drove Einstein crazy, metaphorically speaking. How could a photon travel one route, and also both routes? Or travel both routes, yet only one route? It was ridiculous! "God doesn't play dice," he told his friend and friendly competitor Niels Bohr. "Einstein," Bohr is supposed to have replied, "don't tell God what to do!"[26]

Bohr, notes Wheeler, insisted that his friend Einstein was just looking at the whole thing the wrong way. "We are dealing with two different experiments," Wheeler explains. When the second half-silvered mirror is removed, we have an experiment that tells us which route the photon takes. When mirror H-S2 is put in its location just in front of the two detectors, we have an experiment that tells us the photon traveled both routes at once. "But it is impossible to do both experiments at once," adds Wheeler. "One can observe one feature of nature, or the complementary feature . . . but not both features simultaneously."[27]

Just as importantly, Wheeler adds, this kind of experiment tells us that "what we choose to measure has an irretrievable consequence for what we will find."[28] One vivid way of showing the reality of what Wheeler has called our "observer-participancy" in the realms of quantum reality and "normal reality" is a version of the beam splitter experiment called the delayed-choice experiment.

It's simple: Instead of placing the final half-silvered mirror in the apparatus before we start the experiment, we put it in or take it out at the very last picosecond, *after* the photon *has completed its journey*. Moving the mirror in or out of the experimental setup thus has the effect of influencing what we can say about what has already happened to the photon. In a sense, we affect the photon's *past history*. Wheeler has written about one spectacular version of the delayed-choice experiment that shows how our actions in the here-and-now can even have an effect on what we can say about the universe billions of years ago.

Alternate Reality: Quantum Astronomy

For many years now astronomers have been studying a phenomenon first predicted by Einstein's general theory of relativity called gravitational lensing. General relativity treats gravity as a kind of bending in the fabric of space-time. Highly massive objects like stars and galaxies warp the space-time around them more deeply than do objects with little mass, like planets, asteroids, people, and dust particles. Wheeler has encapsulated this fact of physics in his pithy statement, "Space tells matter how to move, . . . and matter tells space how to curve."[29] Light beams follow paths in space-time, and if space-time is bent then the light beam's path bends, too. Not long after Einstein published his general theory of relativity, astronomers tested this light-bending prediction. They carefully measured the positions of stars near the Sun during a total eclipse, and then compared them with their positions in the sky when the Sun was not in the field of view. Sure enough, the stars' positions were different during the eclipse. The Sun's gravity had bent the light coming from them and caused them to appear to be in places they were not.

Astronomers later realized that objects like galaxies, consisting of hundreds of billions of stars, might also act as gravitational lenses. And, in fact, many cases of cosmic gravitational lensing have now been detected. One example is twin quasars known as 0957 + 561 A and B. Quasars are highly energetic objects with the masses of galaxies, lying billions of light-years distant. They are probably very young galaxies that are pumping out prodigious amounts of energy as supermassive black holes at their centers swallow up gargantuan gobs of matter. Several astronomers in 1979 discovered that these two quasars are actually one. The second is an image of the first caused by the gravitational lensing of a galaxy lying about a quarter of the way between us and the quasar. Two light beams carrying the quasar's image had spread apart by about 50,000 light-years when the second grazed the edge of the intervening galaxy. That galaxy's mass so warped the space-time around it that the second light beam from the quasar was bent around the galaxy. The second light beam from the quasar was thus sent on a path that intersects us. The result? Should Jerry Nelson and his associates point the Keck telescope in the right direction, they will see *two* quasars, 0957 + 561 A and B, where actually only one exists.

Suppose, said Wheeler, we use the case of two images of the same quasar as part of a delayed-choice experiment. The two beams of light from the quasar follow paths 1a and 1b; 1b is the path of the light beam bent by the gravitational lens. We set up a special delayed-choice apparatus in the W. H. Keck Observatory. It has a filter that will allow through only wavelengths of light suitable for an interference experiment, just in case that's what we should choose to do. Also, the filter allows photons to enter the apparatus at only a very low rate. Perhaps we will detect only one photon an hour. Once the two light beams pass through the filters, we label their paths as 2a and 2b. Next we need a lens to focus the light beams and send them along paths 3a and 3b. The light beams will then enter the openings of two fiber-optic cables, which will carry the light through them, no matter how we bend or loop the cables. (Many phone companies now use fiber-optic cables to carry telephone signals on beams of light.) Now the light beams follow paths 4a and 4b in the fiber-optic cables. One fiber-optic cable is longer than the other. Its length is such that we can bring both beams together with the same, or nearly the same, wavelength phase. One beam can enter detector I, the other can enter detector II. Finally, we can place a half-silvered mirror at the junction where the two beams cross, if we wish, and conduct the interference experiment.

Now "we get up in the morning," Wheeler writes, "and spend the day meditating on whether to observe 'which route' or to observe interference between 'both routes.' "[30] Night arrives, we get up from our room in the dorms halfway up the slope of Mauna Kea, and head up to the white dome of the W. M. Keck Observatory. We either place the half-silvered mirror in the apparatus or we do not, depending on our choice. We point the Keck telescope and its mosaicked insect-eye mirror at quasar 0957 + 561 A and B. The light beams from each image of the quasar enter the special delayed-choice apparatus. They pass through the filters and the lens and enter their respective fiber-optics cables. If we have chosen to leave the half-silvered mirror out of the apparatus, we'll observe by which route each photon entered the apparatus. Detector I may go off, or detector II, as each photon makes it through the filter and arrives. If we put the half-silvered mirror in the apparatus, only detector I will fire off as each photon arrives and wave interference takes place.

Now, here's the spooky part. That photon had already gone past the lensing galaxy *billions of years ago*, long before we decided what to do with the half-silvered mirror. So, how did the photon "know" what to do—to follow one route and act like a particle, or follow both at once and act like a wave?

According to Wheeler, following in the footsteps of his mentor Niels Bohr, our perception of what's happening is throwing us off. "It is wrong to speak of the 'route' of the photon" as it travels from quasar through intergalactic space and into the beam-splitter. It is even "wrong" to attribute some kind of "tangibility," as Wheeler puts it, to the photon itself![31] The photon's "reality" is not the kind we are used to in our everyday lives. The "path" it follows is not the kind of path we follow from home to work, from work to lunch, from Los Angeles to San Francisco. "An elementary phenomenon is not a phenomenon until it is an observed phenomenon," Wheeler reminds us; it ain't *nothin'* till we calls it.

And he reminds us of the role *we* play. "Nature at the quantum level is not a machine that goes its inexorable way," Wheeler says. "Instead, what answer we get depends on the question we put, the experiment we arrange, the registering device we choose. We are inescapably involved in bringing about that which appears to be happening."[32]

The "Twenty Questions" of Alternate Reality

Wheeler has offered still another way of trying to map the landscape of quantum reality. It helps explain our role as "observer-participants" in the creation of what is real and gives us a look at another way to understand the Copenhagen Interpretation of quantum mechanics. Wheeler calls his explanation the "surprise version" of the game Twenty Questions. Richard Feynman had this version played on him during graduate school, and Wheeler, his thesis advisor, has been fascinated by it ever since.

In the "normal" version of Twenty Questions, one person leaves the room while the rest of the group decides on a particular person, place, or thing. Then the person who is "it" returns to the room and tries to guess what the person, place, or thing is. The player does this by asking questions. He or she can ask only twenty questions, and they must be of the "yes" or "no" variety. By asking the questions and getting "yes" or "no" answers, the player narrows down the range of possible solutions until he or she finally either knows the solution or makes an educated guess. For example, my friends and I decide that the solution to Twenty Questions is "a cat." You come back to the room and begin asking questions:

- "Is it an animal?" "Yes."
- "Is the animal an extinct animal?" "No." (This question eliminates all dinosaurs, for example.)
- "OK. Hmmm. Is this animal as large as a human?" "No." (Bad question. The animal could be larger or smaller; you still don't know! You have wasted a question.)
- "Is the animal smaller than an adult human?" "Yes."
- "Is this animal a wild animal?" "No." (Excellent question!)
- "Aha! Is this animal a common household pet?" "Damn. Yes."
- "Is this animal a *dog*?!" "No."
- "I think I know the solution. Is this animal a cat?" "Yes."

Pretty simple. Now, though, consider Wheeler's "surprise variation" of Twenty Questions. This time you leave the room and wait . . . and wait . . . and wait. Finally your friends, laughing, tell you to come back in. You return and start asking questions. At first the answers come quickly: "No." "No." "Yes." "No." As you continue to ask questions, though, the pauses become longer and longer before you get a "yes" or "no." Finally, with question number 20, you ask, "Is the answer germanium, the element used in transistors?" "Yes!" Your respondent replies, and the whole group falls to the floor laughing.

What's going on? Well, while you were out of the room the group *did not pick anything at all as the solution to Twenty Questions.* Instead, they decided to allow your questions *to determine the nature of the solution*! The only rule they had was that each person must have some real object in mind as they answered "yes" or "no," and that the object must also match all the "yes's" and "no's" that had been given previously. No wonder the responses got slower and slower!

This is what the land of the quantum is like, according to Wheeler. This is how the Copenhagen Interpretation works. The answer, germanium, did not exist at the beginning of this game of Twenty Questions. All that existed was the observer (the person who is "It" and asking the questions) and the observations (the answers to the questions). The observations created the observed entity, the solution to the game.

In the same way, the measurements we carry out, the observations we make, says Wheeler, play an important role in what we can say about the past of space and time. Our choices make a difference. The consequences of our choices can even ripple backward in space-time, as Wheeler's delayed-choice experiment with the quasars shows. Our choices can even stretch back to the very beginning of the universe. The COBE team and NASA headquarters made a specific set of choices about the instruments COBE could carry. What they have seen, the ripples in the fabric of the cosmos at the very beginning, was in part determined by the choices they made in instrumentation for COBE. Were the ripples already there before COBE detected them? Before Smoot and his colleagues called up the data on computer monitors? Wheeler would say: Those questions are meaningless. Do the measurement.

"Useful as it is under everyday circumstances to say that the world exists 'out there' independent of us, that view can no longer be upheld," Wheeler writes.[33] "Quantum mechanics promotes the mere 'observer of reality' to the 'participator in the defining of reality.' It demolishes the view that the universe exists 'out there.'"[34]

Visions of Reality

What, then, is the nature of reality? Wheeler suggests we consider an old Jewish legend, a folk tale, or *midrash*.[35] In this story Abraham and Jehovah are talking together. Jehovah says to Abraham, "You would not even exist if it weren't for me." To which Abraham replies, "Yes, Lord, that is true. But no one would know you if it weren't for me."

Today, says Wheeler, the two participants in this dialogue have changed. They are humanity and the universe. The universe reminds us, "You would not exist were it not for me. Before the Big Bang nothing existed, and if I ever cease to exist so will everything, including you. How unimportant you are!" And we reply, "Yes, it is true that no *before* existed before the Big Bang, and no *after* will exist after you cease to exist. It is true that without you we could not have come into existence. But remember, O Universe. You consist of phenomena, and every phenomenon requires an act of observation and an irreversible act of amplification. So how would *you* exist without elementary acts of observation such as those we make? Maybe we're not as unimportant as some have suggested."

Exactly how ethereal, how spooky, is that tree outside my office window? How illusory is the redness of the berries still hanging from its twigs? The light reflected off those berries is made of photons, quanta, collapsed quantum wave functions that existed in the strange realm of quantum reality until I just glanced over there. How unreal is the unreality of quantum reality? No matter what the nature of the quantum landscape, those berries are still beautiful to behold.

For Wordsworth and the Romantics of nearly two centuries ago, the phenomena of nature were the ultimate physical realities. They believed that our "eye[s] [are] busy in the distance, shaping things / Which [make our hearts] beat quick." "The harvest of the quiet eye," the inner eye of the mind, is the ultimate reality: the reality of imagination and poetry.

Wheeler says Wait. We are far from even beginning to have an answer to the question What is reality? Do untold billions of acts of observer-participancy weave the foundation of reality? The possibility that this is somehow "true" is enormously intriguing. As we reach the end of the twentieth century, we are glimpsing at least the possibility that the science of quantum physics offers a new understanding of how we collectively create our alternate realities.

In his fascinating essay "Law without Law," Wheeler reminds us that the sciences of astronomy, cosmology, and quantum physics now bring us face to face with one of the most basic questions about the nature of reality. Whether it be the Big Bang or the great inflation or something else, there is little doubt that the universe had a beginning. Space and time themselves began with the Big Bang, as did the four great forces that govern all of the cosmos. Moreover, all the laws of the universe also began at the beginning. And that leads Wheeler to ask the $64,000 question: How?

The universe must have had a way to come into being out of nothingness. This is not the nothingness of the vacuum, of empty space, he reminds us. Space is not empty. At the most intimate and most basic level, in fact, quantum mechanics reveals that space is seething. "On an atomic scale," Wheeler writes, space-time "appears flat, as does the ocean to an aviator far above. The closer the approach, the greater the degree of irregularity." Suppose we were to look at space-time itself from a very close distance, at about 1.6×10^{-33} centimeters (a distance called the Planck length in honor of Max Planck). Then, says Wheeler, "the character of space undergoes an essential change. . . . Multiple connectedness develops, as it does on the surface of the ocean where waves are breaking."[36] Space-time looks like beer foam at this distance. Subatomic particles appear and disappear in an instant.

So when Wheeler says nothingness, he means *nothingness*. No structure. No plan. No law.

But, here we are. We observe and thus participate in the universe. Our observations even affect the nature of the cosmos billions of years ago, as Wheeler's delayed-choice experiment with the quasar shows. "No elementary phenomenon is a phenomenon until it is an observed phenomenon," says Bohr.

The universe must have a way to come into being. This, perhaps, is the ultimate connection between the reality of the quantum and the reality of the cosmos of galaxies and cosmology; of birds and bees and roses and humans; of our observations of the night sky.

"The universe is a self-excited circuit," Wheeler suggests.[37] It begins with the Big Bang, whose earliest fingerprints have been lifted with the sensitive instruments aboard the COBE satellite. The universe expands and cools. Energy decouples from matter and there is light. Hydrogen and helium gas clouds cool, collapse, and form protogalaxies and galactic clusters. The gas collapses still more and breaks up into a myriad contracting whirlpools that shrink, collapse, and light up with fusion fire.

The first galaxies glow with light and at the centers of many are the dark places where gravity rules over all. The black holes suck in all matter around them, and the matter releases huge gouts of energy as it falls into the unfathomable depths. The universe continues to expand and the light from the nascent quasars streams out into space, stretched redder and redder as the eons pass.

Stars explode as supernovas and scatter their heavy, cooked remains across the darkness. New clouds of gas and dust, seeded with heavier elements like carbon and oxygen, collapse into stars. At least one such star is circled by smaller bodies: planets. The third one out, at first hot from its own contraction and formation, cools and contracts. Water boils out of its rocks, steams into the atmosphere, falls as rain, evaporates, and falls again and again.

Hundreds of millions of years pass; the land cools, the water stays liquid and pools into oceans and lakes and rivers and puddles. On the border between land and water, in the mudflats and estuaries, the rigorous laws of chemical combination create life. The simple molecular creatures live and die, grow, change and change again. Less than a billion years has passed since Earth became solid, and already life clings to its watery surface. Life evolves; simple cells become complex clusters of cells. More and more complex lifeforms develop increasingly sophisticated sensory apparatus to observe and participate in the cosmos. Light-sensitive spots become spherical cavities lined with specialized cells; the cavities become simple eyes; the eyes become complex.

More than 4.5 billion years have passed since Earth came into being. More than 15 billion years have elapsed since the universe itself came into existence. It has given birth to observer-participancy. And, says Wheeler, observer-participancy "in turn give[s] tangible 'reality' to the universe not only now but back to the beginning." A self-excited circuit: The universe creates its observer-participants, who in turn give it reality.

And to call the universe a self-excited circuit, adds Wheeler, "is to imply once more a participatory universe."[38]

Wheeler has recently begun referring to his vision of the nature of reality as " 'it' from 'bit.' "[39] In the early 1960s Wheeler began noting interesting parallels between quantum physics and information theory. Claude Shannon of the Bell Laboratories first began formulating information theory in 1948. Information theory is based, essentially, on the development of binary logic, the same principles that make possible the digital computer. The essential unit of information theory is the bit, the binary unit. It represents one of two choices: up or down; black or white; heads or tails; on or off; yes or no.

Shannon and others also showed that information theory offers physicists a new way of looking at entropy. Entropy is basically the randomness in any particular physical system. The more you shuffle a deck of cards, for example, the greater the deck's entropy. Shannon noted that the information a system contains is a function of its entropy. A system's information is the sum total of all its possible messages. So the more one increases, the more the other does also.

Wheeler saw parallels between these aspects of information theory and quantum mechanics. The bit is the "quantum" of information theory. The quantum wave function of quantum mechanics is collapsed by an observer essentially answering one of the surprise "Twenty Questions" with a "yes" or a "no." Moreover, entropy in information theory is tied to the observer's state of mind. The more ignorant I am about the potential amount of information in a system, the greater the number of potential messages the system contains—and the greater the system's entropy. When I *know* what the message is, then the system's amount of information suddenly "collapses." All the other potential bits of information "disappear" and only the one remains. At the same instant, the system's entropy also decreases. It's the same as collapsing the quantum wave function with an irretrievable act of observation and creating a bit of reality from potentiality.

"It" from "bit." Every "it," Wheeler says, be it a particle or field of force or even space-time itself, derives its very existence from answers to yes-or-no questions.

And we are the ones who ask the questions, determine the answers, and together create reality.

5

Alternate Realities: Seeing Things Invisible

Where Alph, the sacred river, ran
Through caverns measureless to man
Down to a sunless sea.

—Samuel Taylor Coleridge
Kubla Khan, or A Vision in a Dream. A Fragment

The great-grandparents of most European and American adults alive today were adults a hundred years ago. What was their world like? In particular, what were the prevailing visions of reality in astronomy, cosmology, and physics for the educated adult layperson in 1897?

Alternate Realities: Then . . .

The 1890s were an exciting time for astronomy. In 1891, for example, George Ellery Hale invented the spectrohelioscope. The spectroscope was already a major instrument for astronomical discoveries. Hale's new instrument made it possible to take a photograph of the Sun with the light from just a single spectral line. The following year E. E. Barnard discovered that a nova emits a cloud of gas as it brightens. It was the first clear evidence that novae are exploding stars.

In 1894 Percival Lowell established his observatory near Flagstaff, Arizona, and began searching for the still-hypothetical ninth planet of the solar system. The search for Pluto would last more than thirty-five years before success was had.

Two years later German astronomer John Schaeberle detected a dim stellar companion to the star Procyon. It was the second white dwarf star to be discovered (the first was the companion to

Sirius), and it suggested that dwarf stars might be common in the Galaxy.

Astronomy also saw significant advances in 1897. Hale set up Yerkes Observatory in Williams Bay, Wisconsin. The Yerkes telescope had a primary lens that was 1 meter in diameter. It was (and still is) the largest refracting telescope on earth.

Astronomers were gathering so much knowledge about stars with spectroscopes that they could classify them by what they learned. In 1897, an astronomer named Antonia Maury modified and refined the spectral classification of stars by using the sharpness of the lines in their spectrum as a guide. That same year, Henry Rowland used a special concave diffraction grating to photograph the sun. He produced a photo of the Sun's spectrum that was 20 meters long.

One of the most important astronomical discoveries of the nineteenth century took place in 1897. On August 12 of that year, the Russian-American astronomer Otto Struve discovered thin clouds of gas and dust lying between the stars. Struve was a man with imagination. The presence of gas and dust in interstellar space was more than just an interesting piece of information. He realized that these clouds could be the source of material for the creation of new stars and planets, and suggested a way for that to happen.

These and other astronomical advances were interesting and at times exciting. The large daily newspapers like the New York *Sun* often carried stories about new astronomical discoveries. Many had heard about Hale and his new telescope in Wisconsin, and Percival Lowell was already somewhat famous. Educated laypeople knew that the Sun was one of many millions of stars in the Milky Way Galaxy. They understood that our Sun and solar system lay somewhere near the middle of the Galaxy, that it was quite large, and that the Milky Way Galaxy filled the entire universe. In fact, the Galaxy *was* the entire universe.

It was a big universe, yes, but it was a *known* universe. We knew who we were and where we were. We knew who our neighbors were and how far away they lived. Many of us assumed there was life on other planets like Venus and Mars.

Problems in astronomy? Yes, some nagging ones, but the average person was probably not aware of them. The biggest headache was the matter of the spiral nebulae. Many nebulae populated the heavens. They were obviously all part of the Milky Way, since *everything* was part of the Milky Way. Some nebulae were proven to be gas clouds. When astronomers turned their big telescopes on others, they turned out to be clusters of stars. But the spiral nebulae stubbornly resisted classification. The spectroscope proved they were made of stars, but the best optical telescopes could still not resolve those stars into visibility. That was the problem. If the spiral nebulae were made of stars, they must be extremely far away. But how far? No one yet knew.

Cosmology? Cosmology as a science really didn't exist in 1897. Two main cosmological visions prevailed among most people in Western culture at the end of the nineteenth century. One was religious and the other was scientific.

Based on the Bible, the religious cosmological vision was relatively simple. Essentially, this vision held, God created the universe at a specific moment in time. It took God six days to create the universe, and on the seventh He rested. Those who held this cosmological vision were not necessarily ignorant or anti-science. Many well-educated people of nineteenth-century America and Europe, people who believed in progress and the march of science, stoutly held to this cosmological vision. It was rooted deeply in their psyche, along with the equally powerful guidance and support of their religious belief systems. It was true that scientists were discovering remains of creatures that obviously had lived many eons ago. Some had even found skeletons that resembled humans. Darwin's *On the Origin of Species* had been published in 1859 and was still the subject of great controversy. For all that, the religious cosmological vision still prevailed for many people.

Others, however, had taken to heart and soul the scientific cosmological vision of the late nineteenth century. This second common belief about the origin of the universe was that it *had* no beginning. The universe was very large and it was essentially static. Newton himself believed this to be true, and had said so in his writings. His belief had in large part been based on his theological leanings, but it was his position nonetheless.

So, the universe had always existed. It had always been the size that it was—big, and consisting of the Milky Way and its stars, planets, and nebulae. Change was real, however. Evolution was real, and humans had evolved from more primitive ancestors. Giant creatures had once walked the earth, and perhaps our ancestors had battled with them for survival.

The 1890s were exciting times for physics, as for astronomy. In 1894, for example, J. J. Thomson found that the velocity of cathode rays was much slower than that of light, which meant that cathode rays could not be a form of electromagnetic waves. Much more momentous was the discovery of X rays the following year by Wilhelm Roentgen. Within weeks of his announcement, doctors were using the new rays to peer inside the bodies of living humans. That same year, the Scottish physicist Charles Wilson invented the cloud chamber. This device made it possible to detect the paths of invisible charged particles as they passed through the chamber. The cloud chamber would eventually become a powerful tool for twentieth-century particle physics.

In 1897, the mystery of cathode rays was solved. J. J. Thomson determined that they were tiny particles with a negative electric charge. Today we know them as electrons.

Despite these breakthroughs, in 1897 the vision of reality offered by physics was in ferment. Two great problems confronted physics: the ether and the ultraviolet catastrophe.

Michelson and Morley's experiments, begun in 1888, had completely failed to detect the presence of the all-pervasive ether. If light was a wave, as Thomas Young had proven, it needed a medium in which to travel. That medium was the ether. But, either there was no ether or the earth was unmoving. Neither could be true.

In 1892 Fitzgerald and Lorentz had offered their Lorentz–Fitzgerald contraction formula to explain away the failure of the Michelson–Morley experiments. Objects, they said, contracted in the direction that they were moving. That would cancel out any changes in the speed of light caused by Earth's movement through the ether and produce the findings of Michelson and Morley. But few people found this a satisfying explanation.

Speaking of light: the ultraviolet catastrophe was another perplexing conundrum. The prevailing theories of atoms and electromagnetic radiation said that a blackbody, when heated up, should begin to emit large amounts of blue and violet light and ultraviolet radiation. In reality, though, this did not happen. But if the theories were askew, then the very models of physical reality must be drastically wrong.

Here, too, some researchers had tried to come up with patchwork solutions. Wilhelm Wein and Lord Rayleigh had each devised formulas that explained away some of the shortcomings in the prevailing theory of blackbody radiation. But these were not enough. The catastrophe loomed, and in 1897 Max Planck's imaginative solution was still three years away.

Alternate Realities: Now . . .

Today's visions of reality are considerably different from the ones of our great-grandparents in 1897. The vision of the universe offered by astronomy and cosmology back then was exciting but still somewhat "safe" and comfortable. The universe was an orderly place, governed by Newton's laws of motion and the universal law of gravity. It was an astronomical and cosmological landscape similar to the social and cultural landscape of the times. Most educated laypeople in Europe and America still knew their neighbors. They rarely traveled more than a few dozen kilometers from home. They knew their place in the world and worked hard to make it better for their children.

By the latter part of the twentieth century, the visions had changed.

Then the cosmological vision most familiar to the educated layperson was of an expanding universe that was 15 billion years old. It was not only very big, it was also very empty and very lonely. The Moon had been visited and found to be dead. Martian meteorites held possible evidence of fossil life. Perhaps life had once existed on that planet, but it was now extinct—a chilling promise of what is to come for us?

Then Europa, a moon of Jupiter first discovered by Galileo in that *annus mirabilis* 1609, became an object of hope. In the early and mid-1980s the Voyager space probes had first discovered that Europa was completely covered with ice and that the icy surface was criss-crossed with dark brown lines.[1] Could these lines be crevasses in the ice? Could the brown color be caused by organic compounds in water welling up from beneath? Scientists, science journalists, and science fiction writers began speculating about the possibility of oceans of liquid water beneath Europa's icy crust.[2] And liquid water might mean life. The close-up photos from the aptly named Galileo space probe of more than a decade later seemed to confirm the speculations; NASA began talking about sending landers to Europa to search for life.

Most people who think about life beyond Earth, however, dream larger dreams. They hope that the science fiction visions of *Star Trek* are true and that we are not alone in the universe, that other sentient beings roam the spaceways, and that we will someday meet them—if we haven't already. But television shows and movies to the contrary, we're still waiting for someone to stop by and say hello.

The alternative reality offered by astronomy and cosmology may seem cold and pessimistic, particularly when the speculations about life on Europa, or Vulcans, or UFOs are disregarded. But in a society that seems beset with random violence and moral breakdown, the visions of contemporary cosmology may offer many an alternate reality that is a compelling diversion from the chaos of daily life.

And then, there's quantum physics.

The images of reality conjured up by astronomy and cosmology have already penetrated deeply into the "collective unconscious" of Western society. The same cannot yet be said for the oft-times bizarre implications of quantum mechanics. But just as the astronomical and cosmological discoveries of Edwin Hubble in the 1920s and Arno Penzias and Robert Wilson in the 1960s eventually permeated the inner visions of the average nonscientist, so will the discoveries and explanations of people like John Wheeler.

This new vision of quantum reality is strange, but it is in some ways profoundly intimate. Contemporary cosmology tells us that the universe had a beginning. Quantum cosmology adds that, in the beginning, all the matter and energy in the universe existed for one instant in utterly close association. That association still prevails. Experiments carried out in the 1980s have shown the reality of what physicists call "nonlocality." By that they mean there is now evidence of an ongoing intimate connection between every subatomic particle in the cosmos. Under the right conditions, it can be shown that a particle at one end of the universe will "know" what another particle is doing, even though they cannot possibly communicate with one another. This "knowing" happens faster than light can carry the information from one to the other.

In other words, in this emerging vision of reality offered by quantum physics, everything is connected to everything else. Not just metaphorically, or theologically, but in some mysterious but "real" way.

William Wordsworth and his Romantic colleagues would have loved it. These were the poets, painters, philosophers—and scientists—who firmly believed that all of nature was a single entity imbued with Spirit. These were the people whose "harvest of the quiet eye" was a vision of reality imbued with life and meaning. Now, nearly two centuries after their time, the "inner eye" that in tranquillity and memory reshapes sensory experiences into new realities has met quantum reality.

But there is even more to this alternate reality, and more that the Romantic poets would appreciate. For the universe of John Wheeler is an observer-participatory universe. It is a cosmos in which we are not minuscule motes in a meaningless vision, but active participants in the creation of reality itself. *We* are the ones who, to paraphrase Wheeler, first establish the iron posts of observation and then weave the brilliant tapestry of reality between them.

It is a cosmos where the increase of knowledge, of answers, and of wisdom truly decreases chaos and increases order. It is a cosmos that rewards asking questions.

And who asks the questions in an observer-participatory universe? The poets, the painters, and the writers; the philosophers and the scientists; the dreamers; also the plumbers, the teachers, and the farmers. The poor and the rich; the women and the men; the old and, especially, the young.

We ask the questions. We spin the web. We are the ones with "eyes" that are busy in the distance, "shaping things" with our raw observations, collapsing quantum wave functions from probability to reality. We are Coleridge's dreamers in "Kubla Khan." We have "drunk the milk of Paradise," and we see things invisible.

Appendix

Scientific Nomenclature and Conversion Table

Whenever this book refers to scientific measurements it uses the International System of Units (*Système internationale d'unités*, abbreviated SI) or so-called metric system. This is the most common measurement system used in the world today, and the one officially adopted by the scientific community. SI units are in powers of ten. The prefix for the name of the unit indicates which power of ten to use. I've chosen to use metric system nomenclature throughout the book as if it were the accepted measurement system everywhere in the world, rather than everywhere except the United States and Great Britain. I do resort at times to comparisons with the English of feet and miles, as well as using more concrete comparisons. Those of you who would like to know how to "translate" SI into English and vice versa, however, can refer to the following conversion table and an explanation of the more common SI abbreviations and prefixes.

The Basic SI Units

Quantity	Unit	Symbol
Amount of substance	mole	mol
Electric current	ampere	A
Frequency	hertz	Hz
Length	meter	m
Luminance intensity	candela	cd
Mass	kilogram	kg
Temperature	kelvin	K
Time	second	s

Other Important SI Units

Quantity	Unit	Symbol
Electric potential difference	volt	V
Energy	joule	J
Force	newton	N
Power	watt	W

Scientific Measurement Units and their Equivalents

SI		English	
Length:			
1 angstrom (Å)	= 0.000 000 000 1 m		
1 nanometer (nm)	= 0.000 000 001 m		
1 micrometer (μm)	= 0.000 001 m		
1 millimeter (mm)	= 0.001 m	= 0.03937	inch
1 centimeter (cm)	= 0.01 m	= 0.3937	inch
1 meter (m)		= 39.37	inches
		= 3.28	feet
1 kilometer (km)		= 0.6214	miles
Mass:			
1 milligram (mg)	= 0.001 g		
1 gram (g)		= 0.035	ounces
		= 0.002046	pounds
1 killigram (kg)	= 1,000.0 g	= 2.2046	pounds

Abbreviations and Prefixes for Numbers

Symbol	Prefix			Factor
n	nano	thousand millionth (billionth)	0.000 000 001	10^{-9}
μ	micro	millionth	0.000 001	10^{-6}
m	milli	thousandth	0.001	10^{-3}
c	centi	hundredth	0.01	10^{-2}
k	kilo	thousand	1,000.0	10^{3}
M	mega	million	1,000,000.0	10^{6}
G	giga	thousand million (billion)	1,000,000,000.0	10^{9}
T	tera	million million (trillion)	1,000,000,000,000.0	10^{12}

Some Important Physical Values and Constants in Astronomy and Physics

Gravitational acceleration, Earth (g)	9.806 meters per second per second (m/s^2)
Gravitational constant (G)	6.672×10^{-11} m^3/s^2 kg
Light-year	9.46×10^{12} km
Mass of a hydrogen atom (m_H)	1.6735×10^{-27} kg
Mass of a neutron (m_n)	1.6749×10^{-27} kg
Mass of a proton (m_p)	1.6726×10^{-27} kg = 1,836.1 m_e
Mass of an electron (m_e)	9.10956×10^{-31} kg
Planck constant (h)	6.626×10^{-34} J s
$\hbar$ ($h/2\pi$)	1.054×10^{-34} J s
Planck length	1×10^{-33} cm
Planck time	1×10^{-43} s
Speed of light in a vacuum (c)	299,792 km/s

Notes

I. In a Grain of Sand

1. Feynman shared the prize with a contemporary, physicist Julian Schwinger. In fact, the two men had both attended the same high school as teenagers.
2. For a lucid and detailed retelling of Feynman's work on QED, see James Gleick, *Genius: The Life and Science of Richard Feynman* (New York: Pantheon Books, 1992).
3. Reproduced in Michael Gamer, "William Wordsworth (1770–1850) Samuel Taylor Coleridge (1772–1834)" (http://www.english.upenn.edu/~mgamer/Romantic/lbprose.html#preface).
4. Classicism is a literary and artistic movement that stressed clearness and elegance through attention to traditional artistic forms, particularly Greek and Roman models. English writers like Francis Bacon and Ben Jonson are good examples of classicists, and the movement reached its peak with the work of Alexander Pope.
5. A couple of centuries later, the British author Sir John Squire wrote the following response to Pope's lines:

   "It did not last: the Devil, howling *Ho*!
   Let Einstein be! restored the status quo."

   Of course, Einstein didn't exactly "restore the status quo." From *The Columbia Dictionary of Quotations* (licensed from Columbia University Press. Copyright © 1993 by Columbia University Press. All rights reserved.), in *Microsoft Bookshelf '95* (Redmond, WA: Microsoft Corp., 1995).

6. This excerpt is from the text of *The Two-Part Prelude* reproduced in Jonathan Wordsworth, ed., *William Wordsworth: The Pedlar, Tintern Abbey, The Two-Part Prelude* (London: Cambridge, 1985).
7. Duncan Wu, "Tautology and Imaginative Vision in Wordsworth." *Romanticism on the Net: An Electronic Journal Devoted to Romantic Studies* (http://www-sul.stanford.edu/romnet/tautology.html).
8. The *sensorium* is collectively the entire sensory apparatus, the parts of the brain—and by extension, the mind—concerned with the reception and interpretation of sensory stimuli.
9. In *Some Poems from* Complete Poetical Works (1888). *Project Bartleby, Columbia University* (http://www.cc.columbia.edu/acis/bartleby/wordsworth/ww138.html).
10. In Marj Tiefert, editor, *Hypertext Poems from the Coleridge Archive* (http://www.lib.virginia.edu/etext/stc/Coleridge/poems/Christabel.html).
11. In Marj Tiefert, editor, *Hypertext Poems from the Coleridge Archive* (http://www.lib.virginia.edu/etext/stc/Coleridge/poems/Rime—Ancient—Mariner.html#I).
12. In *The Poetical Works of John Keats. Project Bartleby, Columbia University* (http://www.cc.columbia.edu/acis/bartleby/keats/keats52.html).
13. In *Selected Poetry of George Gordon, Lord Byron (1788–1824). Department of English. University of Toronto* (http://library.utoronto.ca/www/utel/rp/poems/byron10.html)
14. See *The ABC's of the Human Mind* (Pleasantville, NY: The Reader's Digest Association, 1990), 22.
15. See Walter Truett Anderson's book, *Reality Isn't What It Used to Be: Theatrical Politics, Ready-To-Wear Religion, Global Myths, Primitive Chic, and Other Wonders of the Postmodern World* (San Francisco: Harper & Row, 1990), for a fascinating look at how social reality has changed in our time.
16. The quotes from Blake's poems come from *The Digital Blake Project. Department of English. University of Georgia* (http://virtual.park.uga.edu/~wblake/eE.html).
17. Anderson, *Reality Isn't What It Used to Be*, 72.
18. Charles Percy Snow, *The Two Cultures and the Scientific Revolution* (Cambridge, England: Cambridge University Press, 1961). Two years later Snow released an updated version entitled *The Two Cultures: And a Second Look* (Cambridge, England: Cambridge University Press, 1963).

II. Starry Messengers, Eyes of Glass

1. From *The People's Chronology* (licensed from Henry Holt and Company, Inc. Copyright © 1994 by James Trager. All rights reserved.), in *Microsoft Bookshelf '95* (Redmond, WA: Microsoft Corp., 1995).
2. *Webster's Ninth New Collegiate Dictionary*, Frederick C. Mish, editor-in-chief (Springfield, MA: Merriam-Webster, Inc., 1988), 1051.
3. Excerpted from *Compton's Interactive Encyclopedia* (Copyright © 1993, 1994, Compton's NewMedia, Inc.).
4. From *The People's Chronology*, in *Microsoft Bookshelf '95*.
5. From *The Concise Columbia Encyclopedia* (licensed from Columbia University Press. Copyright © 1995 by Columbia University Press. All rights reserved), in *Microsoft Bookshelf '95* (Redmond, WA: Microsoft Corp., 1995).
6. BCE stands for "before the current era." This is the standard abbreviation used by historians and paleontologists, replacing "BC" or "before Christ." "CE," for "current era," is now commonly used instead of "AD," or "anno Domini," Latin for "in the year of Our Lord."
7. Alexander Hellemans and Bryan Bunch, *The Timetables of Science: A Chronology of the Most Important People and Events in the History of Science* (New York: Simon and Schuster, 1988), 1.
8. Hellemans and Bunch, *The Timetables of Science*, 5–9.
9. Because Earth actually wobbles on its axis, the North and South Poles never point at the same location in the sky. Instead, they trace an imaginary circle in the heavens with a period of 28,000 years. Three bright stars have been the Northern Hemisphere pole stars over the last 14,000 years. Vega, a very bright star in the constellation Lyra (the Lyre), was the pole star about 14,000 years ago. In about 2700 BCE the pole star was Thuban, a star in the constellation Draco, the Dragon. The pole star today is Polaris, the last star in the handle of the Little Dipper.
10. According to Eisler and other goddess researchers, most of the goddesses of Greek and Roman mythology are the "psychic descendents" of a much older and singular goddess. The original goddess with her three aspects of maiden, mother, and crone became fragmented in later "patriarchal" mythologies and religions into many different goddesses. For example, the Greek goddess Artemis, the goddess of the hunt, was traditionally associated with marriage and children—a "mother" aspect of the original goddess. Like her Roman counterpart Diana, Artemis was also associated with both the Moon and the Earth. See *The Concise Columbia Encyclopedia*, in *Microsoft Bookshelf '95*.

11. A. Pannekoek, *A History of Astronomy* (New York: Dover, 1961, 1989), 83–84.
12. *Ibid*, 84.
13. The best-known and most popular books by Hawkins are *Stonehenge Decoded* (New York: Doubleday, 1965) and *Beyond Stonehenge* (New York: Harper & Row, 1973). *Stonehenge Decoded* was the book that first brought the astronomical significance of Stonehenge to popular attention. Two of Thom's more important and better-known books on megalithic astronomy are *Megalithic Sites in Britain* (London: Oxford University Press, 1967) and *Megalithic Lunar Observatories* (London: Oxford University Press, 1971). Though less well known than Hawkins, Thom and his books are perhaps more thoroughly researched.
14. A medium's *absolute* refractive index (which scientists symbolize with n) is the ratio of the speed of light in the medium to the speed of light in a vacuum (which is about 299,792 kilometers per second). A medium's *relative* refractive index is the ratio of the speed of light in one medium to that of the adjacent medium.
15. The Platonic solids are the five solid objects whose faces are each a regular polygon (a closed plane whose sides are of equal length and which meet at equal angles). They are the tetrahedron, whose faces are four equilateral triangles; the hexahedron, or cube, with six squares as its faces; the octahedron, with faces of eight equilateral triangles; the dodecahedron, with twelve regular pentagons; and the icosahedron, whose twenty faces are equilateral triangles.
16. The mathematical formula, for those who are interested, is

$$M = F/f$$

where M is the magnifying power of the telescope, F is the focal length of the objective, and f is the focal length of the eyepiece.
17. The most obvious example today of the strength of belief is the system called creationism. A small but vocal minority of fundamentalist Christians believe firmly in the "scientific" accuracy of the biblical creation story. Genesis essentially states that God created the world in "six days," and on the seventh day rested from His labors. It also offers an explanation of the origin of the human species. Creationists insist that these stories are literally true; therefore, all supposed evidence to the contrary is either a trick of the Devil or somehow a test of our willingness to place our unwavering trust in "the Word of God." In fact, creationism is nothing more than a theological belief system based on a narrow reading of the Bible and a willful ignoring of solid scientific data. Biblical scholarship has in the last century become a highly sophisticated field of study. It makes use of linguistics, economics, archaeology, anthropology, history, and even astron-

omy and physics. One result has been a clear understanding of the biblical creation stories as forms of creation myths. We now understand the origins of those specific myths, when they became part of the Jewish scriptures, and what meaning they held for the peoples of those times. At the same time, scientific advances in biology, archaeology, and paleontology have established evolution as a solid scientific theory—not speculation, but well-documented and widely accepted for its evidence. Creationists cling to their belief despite the overwhelming evidence to the contrary.

18. The story of Galileo's "discovery" of Saturn's rings appears in numerous references. These details come from Richard Learner, *Astronomy through the Telescope* (New York: Van Nostrand Reinhold, 1981), 13.
19. In 1621 the Dutch mathematician Willebrord Snell discovered the law that describes the refraction of light through different media, now called Snell's Law. Newton discovered that Snell's Law applied to each color that composed white light, and not just to white light as a whole.
20. For more details on the Hale Telescope at Palomar Observatory and Baade's discoveries about Cepheid variables, see Learner, *Astronomy through the Telescope*, 141–144.
21. William Parsons, the third Earl of Rosse, who built a famous telescope in Ireland called "The Leviathan," was the first astronomer to dabble with the idea of a segmented mirror. In the late 1940s an Italian astronomer named G. Horn-d'Arturo combined sixty-one small hexagonal mirrors into a single 1.8-meter mirror for a reflecting telescope. It produced a good image, but always had to be kept in a vertical position.

III. Ripples of Light

1. Actually, the nonexistant centrifugal force is not really a "force" like gravity or magnetism but rather an artifact of our perception. It seems to us that one body orbiting another is held in orbit by a balance of an inward-pulling "centripetal force" and an outward-pulling "centrifugal force." In fact, only one force is in action, the centripetal force we know as gravity.
2. This quote appears in English translation in Pannekoek's *A History of Astronomy*, 272. Its original publication is in Christiaan Huygens's *Traité de la Lumière, et Discours de la Cause de la Pensanteur* (Leipzig, 1885), 119.
3. In Arthur S. Eddington, *The Nature of the Physical World* (London: Cambridge University Press, 1932), 85.

4. The kelvin is the SI (*Système international d'unités*, or metric system) symbol for thermodynamic temperature. A kelvin is the same as a degree Celsius, the more commonly known metric system temperature measurement. However, degrees Celsius start from the freezing/melting point of water and go up and down from there. The kelvin temperature range begins at absolute zero, which is defined as −273.15 degrees Celsius. Before 1967 scientists expressed a kelvin temperature as "degrees kelvin" (with the symbol °K). Since then, by international agreement, it is simply expressed as "kelvins" or with the symbol K. For more information on the SI, see the appendix.
5. For a detailed exploration of antimatter, the history of its discovery, its role in science fiction novels, and its possible use in our future, see Robert L. Forward and Joel Davis, *Mirror Matter: Pioneering Antimatter Physics* (New York: Wiley, 1988).
6. *The American Heritage® Dictionary of the English Language*, 3rd ed. (New York: Houghton Mifflin, 1992).
7. Much of this information on the genesis of the COBE project came via e-mail from Dr. Charles Bennett, NASA Goddard Space Flight Center, and from John C. Mather and John Boslough, *The Very First Light: The True Inside Story of the Scientific Journey Back to the Dawn of the Universe* (New York: Basic Books, 1996).
8. This part of the COBE story is strongly colored by individual perceptions, personality clashes, and some very disturbing activities. George Smoot's version of the story is recounted in his book *Wrinkles in Time*. However, Mather, the COBE Project Scientist and leader of the entire scientific endeavor for COBE, has a somewhat different story to tell in Mather and Boslough's *The Very First Light*. E-mail from Charles Bennett and Edward Cheng about this part of the story, which was forwarded to the author, supports Mather's recollections. Those readers who wish to get the details of the unfortunate actions of a member of the COBE team are urged to read Mather and Boslaugh's book.
9. This quote is reported by Charles Bennett, e-mail, March 25, 1997.
10. Preliminary reports on the MaCHO discoveries appeared in the sci.astro newsgroup of the Usenet, as carried on the Internet, on September 21, 1993.
11. Charles Bennett, e-mail, March 25, 1997.
12. The quotes come from the Reuters news article as it appeared under the headline "Religion and the Big Bang Discovery," Seattle *Post-Intelligencer*, April 25, 1992, B6.
13. This quote originally appeared in J. C. Mather, "Cosmic Background Explorer (COBE): Report to the Space Science Board" (NASA/Goddard Space Flight Center, Greenbelt, Maryland, 1977), 3.

IV. Quantum Realities

1. John A. Wheeler, "The Computer and the Universe," *International Journal of Theoretical Physics* 21 (1982).
2. Michell was a fine example of the "amateur scientist" so prevalent in the eighteenth and nineteenth centuries. Though an Anglican priest by profession and not a "professional" scientist, Michell made significant contributions to several scientific disciplines. In addition to his treatise on magnets, Michell also wrote a paper in 1760 describing a mechanism for earthquakes. His hypothesis was remarkably accurate for its day. In 1783 he wrote a letter to Henry Cavendish containing the first theoretical description of what we today would call a black hole. For more on this last contribution to astrophysics, see Joel Davis, *Journey to the Center of Our Galaxy* (Chicago: Contemporary Books, 1991, 1992), 131–132.
3. Henry built the first practical electromagnets and electrical motors in 1831, and by 1840 had developed an electrical relay that made the telegraph possible. Samuel Morse, credited with inventing the telegraph, actually worked closely with Henry and incorporated Henry's ideas into his invention.
4. The first decades of the nineteenth century were a golden age for chemistry. For example, in 1803, the same year Dalton proposed his atomic theory of matter, Smithson Tennant discovered iridium and potassium and William Wollaston discovered palladium and rhodium. Louis-Nicolas Vauquelin discovered uric acid in 1811. The same year Amedeo Avogadro proposed that equal volumes of different gases contain equal numbers of particles, today known as Avogadro's Law. Friederich Strohmeyer discovered cadmium in 1817. The prolific chemist Jon Jakob Berzelius discovered silicon in 1823, zirconium in 1824, and titanium in 1825.
5. They also introduced the terms *anode*, *cathode*, *electrode*, *electrolyte*, and *electrolysis*. Whewell, by the way, is the person who coined the term *scientist*. He introduced it in 1833 at a meeting of the British Association for the Advancement of Science and popularized the word in his 1840 book *The Philosophy of the Inductive Sciences*.
6. The charge, said Stoney, had a value of about 10^{-20} coulombs. This turned out to be very close to the actual known value of the electrical charge of an electron, which today is measured at 1.602×10^{-19} coulombs.
7. Planck's constant, h, is measured in units of energy multiplied by time, known as action. Its numerical value is equal to about 6.63×10^{-34} J s (joule-seconds; a joule is the SI unit of work), or about 4.14×10^{-15} eV s (electron volt-seconds). In much of the mathematics of

quantum mechanics, physicists use a variation of Planck's constant symbolized by $\hbar$ (called "h-bar"); $\hbar = h/2\pi$, and is equal to 1.054×10^{-34}J s.

8. Einstein himself made another epochal contribution to physics in 1916. That was the year he published his theory of gravitation, known as the general theory of relativity.
9. Actually, electrons and other subatomic particles don't really "spin." These entities aren't really tiny spheres with well-defined spin axes. "Spin" is a quantum property of entities that describes what a particle "looks like" from different directions. For example, a particle with spin 0 looks the same from any direction. A particle with spin 1 looks the same only after it has been "spun" 360 degrees. A particle with spin 2, on the other hand, looks the same when it makes a turn of only 180 degrees. And a particle with spin $\frac{1}{2}$ must make two complete "revolutions" (720 degrees) before it looks the same as when it started! Today physicists divide all the subatomic particles in the universe into two general categories: those with "spin" of $\frac{1}{2}$ and those with "spin" of 0, 1, and 2. The former includes the protons, neutrons, and electrons that make up atoms, which in turn make up matter. The latter category includes the subatomic particles that carry the forces—electromagnetism, the strong nuclear force, the weak force, and gravity—that act on the former category of particles.
10. Werner Heisenberg, *Physics and Beyond* (New York: Harper & Row, 1971), 38.
11. Max Born, *Atomic Physics* (New York: Hafner, 1957), 102.
12. Heisenberg, *Physics and Beyond*, 76.
13. John A. Wheeler and Wojciech H. Zurek, eds., *Quantum Theory and Measurement* (Princeton, NJ: Princeton University Press, 1983), i.
14. John Archibald Wheeler, "Law without Law," in Wheeler and Zurek, eds., *Quantum Theory and Measurement*, 183–213.
15. The line "Nature, red in tooth and claw" is from Alfred Tennyson's poem "In Memoriam," published in 1850. Tennyson was born in 1809, eleven years after the publication of *Lyrical Ballads*, and died in 1892, the same year Fitzgerald and Lorentz devised their Lorentz–Fitzgerald contraction formula to explain the failure of the Michelson–Morley experiments to detect the "ether."
16. From William Wordsworth's "Lines Composed a Few Miles above Tintern Abbey."
17. John A. Wheeler, *The Frontiers of Time* (Amsterdam: North-Holland, 1979).
18. John A. Wheeler, "Beyond the Black Hole," in *Some Strangeness in Proportion*, ed. Harry Woolf (Reading, MA: Addison-Wesley, 1980).

19. Quoted in Jeremy Bernstein, *Quantum Profiles* (Princeton, NJ: Princeton University Press, 1991), 96.
20. Berkeley's position has been summarized in the humorous aphorism popular among college students for generations: "I no see you, you no be you."
21. Heisenberg, *Physics and Beyond*, 41.
22. The theological implications of this position are summed up in the following two limericks, quoted in J. C. Polkinghorne, *The Quantum World* (Princeton, NJ: Princeton University Press, 1984), 66:

    There once was a man who said, "God
    Must think it exceedingly odd
    If he finds that this tree
    Continues to be
    When there's no one around in the Quad."

    Dear Sir, Your astonishment's odd;
    I am *always* around in the Quad!
    And that's why the tree
    Will continue to be,
    Since observed by Yours faithfully, God.

23. Fans of the fictional character Commander Data in the TV series "Star Trek: The Next Generation" may take exception to this. So may fans of Isaac Asimov's robot stories, as well as many proponents of the development of artificial intelligence, or AI, in computers.
24. John A. Wheeler, "Our Universe: The Known and the Unknown," *American Scientists* 56 (1968): 1.
25. It's no longer a thought-experiment; various physicists have now conducted different versions of it.
26. Jacob Bronowski, *The Ascent of Man* (Boston: Little, Brown, 1973), 122.
27. Wheeler, "Law without Law," 183.
28. *Ibid.*
29. Charles W. Misner, Kip S. Thorne, and John A. Wheeler, *Gravitation* (San Francisco: Freeman, 1973), 23.
30. Wheeler, "Law without Law," 192.
31. *Ibid*, 184.
32. *Ibid*, 185.
33. *Ibid*, 194.
34. John Wheeler, "The Quantum and the Universe," in *Relativity, Quanta, and Cosmology*, vol. 2, ed. M. Pantaleo and F. de Finis (New York: Johnson Reprint Co., 1979).
35. A *midrash* is a kind of teaching story. Jesus used them frequently, according to the Gospels of the Christian Bible's New Testament.

They are called parables. The Christmas story in the Gospel of Matthew is also a good example of a *midrash*, as used by the Jewish–Christian communities of the early first century CE.

36. John A. Wheeler, "On the Nature of Quantum Geometrodynamics," *Annals of Physics* 2 (1957): 604–614.
37. Wheeler, "Beyond the Black Hole."
38. Wheeler, "Law without Law," 209.
39. John Horgan, "Questioning the 'It' from 'Bit.'" *Scientific American* 264 (June 1991): 37.

V. Alternate Realities: Seeing Things Invisible

1. See Joel Davis, *Flyby: The Interplanetary Odyssey of Voyager 2* (New York: Atheneum, 1987) for details on the discoveries of the Voyager 1 and 2 space probes.
2. The first science journalist to publish on this was probably Richard Hoagland. Today mainly known for his controversial theories about the so-called face on Mars, in the late 1970s and 1980s Hoagland was writing for *Star & Sky* magazine and working as a science consultant to CNN News. Science fiction writer Arthur C. Clarke later incorporated the idea of life on Europa into his novel *2010*, the sequel to the movie (and his book) *2001: A Space Odyssey*.

Bibliography

Anderson, Walter Truett. *Reality Isn't What It Used to Be: Theatrical Politics, Ready-To-Wear Religion, Global Myths, Primitive Chic, and Other Wonders of the Postmodern World*. San Francisco: Harper & Row, 1990.

Asimov, Isaac. "Beyond Light." *Magazine of Fantasy and Science Fiction* (December 1991): 121–130.

Asimov, Isaac. *The Human Brain: Its Capacities and Functions*. Boston: Houghton Mifflin, 1963.

Babcock, Horace W. "Adaptive Optics Revisited." *Science* (July 20, 1990): 253–257.

Barlow, Robert B. Jr., "What the Brain Tells the Eye." *Scientific American* (April 1990): 90–95.

Barrow, John, and Joseph Silk. *The Left Hand of Creation: The Origin and Evolution of the Expanding Universe*. New York: Basic Books, 1983.

Begley, Sharon. "Heavens!" *Newsweek* (June 3, 1991): 47–52.

———. "Mapping the Brain." *Newsweek* (April 20, 1992): 66–70.

Berman, Morris. *Coming to Our Senses: Body and Spirit in the Hidden History of the West*. New York: Bantam Books, 1989.

Bernstein, Jeremy. *Quantum Profiles*. Princeton, NJ: Princeton University Press, 1991.

Bernstein, Jeremy, and Gerald Feinberg, eds. *Cosmological Constants: Papers in Modern Cosmology*. New York: Columbia University Press, 1986.

Berry, Richard. "Seeing Sharp: A Revolutionary New Ground-Based Telescope Can Take Extraordinarily Sharp Pictures of Celestial Objects." *Astronomy* (July 1990): 38–39.

Bethe, Hans A. "Supernovae." *Physics Today* (September 1990): 24–27.

Born, Max. *Atomic Physics*. New York: Hafner, 1957.

Boslough, John. "Inside the Mind of John Wheeler." Reader's Digest (September 1986): 106–110.

Bronowski, Jacob. *The Ascent of Man*. Boston: Little, Brown, 1973.

Brush, Stephen G. "How Cosmology Became a Science." *Scientific American* (August 1992): 62-68.

Caelli, Terry. *Visual Perception Theory and Practice*. New York: Pergamon Press, 1981.

Campbell, Joseph. *Historical Atlas of World Mythology: Volume I. The Way of the Animal Powers*. New York: Harper & Row, 1983.

Campbell, Joseph, and Bill Moyers. *The Power of Myth*. New York: Doubleday, 1988.

Cannon, Bill. "God's Magnifying Glass: Dark Matter Reveals Hidden Universe." Press Release. Seattle: University of Washington, June 7, 1993.

Carpenter, Betsy. "Eye in the Sky." *U.S. News & World Report* (December 17, 1990): 36.

Casimir, Hendrik B. G. *Haphazard Reality: Half a Century of Science*. New York: Harper & Row, 1983.

Chaikin, Andrew. "The Ultimate Time Machine." *Popular Science* (March 1992): 68–72, 82–83.

Chanowitz, Michael S. "The Z Boson." *Science* (July 6, 1990): 36–42.

Christianson, Gale E. *This Wild Abyss: The Story of the Men Who Made Modern Astronomy*. New York: Free Press, 1978.

Cohen, Bernard I. *Revolutions in Science*. Cambridge, MA: Harvard University Press, 1985.

Cornell, James. *The First Stargazers*. New York: Charles Scribner's Sons, 1981.

Corsi, Giovanna, et al., eds. *Bridging the Gap: Philosophy, Mathematics, and Physics*. Boston: Kluwer Academic Publishers, 1993.

Cowen, Ron. "Cosmological Controversy: Inflation, Texture, and Waves." *Science News* (May 22, 1993): 328–330.

Crease, Robert P., and Charles C. Mann. *The Second Creation: Makers of the Revolution in 20th-Century Physics*. New York: Macmillan, 1986.

D'Abro, A. *The Rise of the New Physics*. 2 vols. New York: Dover, 1951.

Davies, Paul. "What Hath COBE Wrought?" *Sky & Telescope* (January 1993): 4–5.

Davies, Paul, and John Gribbin. *The Matter Myth: Dramatic Discoveries That Challenge Our Understanding of Physical Reality*. New York: Simon and Schuster, 1992.

Davis, Joel. *Journey to the Center of Our Galaxy*. Chicago: Contemporary Books, 1991.

———. *Mapping the Mind: The Mysteries of the Human Brain and How It Works*. New York: Carol Publishing Group, 1997.

Davis, Marc, F. J. Summers, and David Schlegel. "Large-Scale Structure in a Universe with Mixed Hot and Cold Dark Matter." *Nature* (October 1, 1992): 393–398.

Davis, Michael. *William Blake: A New Kind of Man*. Berkeley: University of California Press, 1977.

"Dazzling Views from Europe's NTT." *Sky & Telescope* (June 1990): 596–599.

D'Espagnat, Bernard. *Reality and the Physicist: Knowledge, Duration, and the Quantum World*. New York: Cambridge University Press, 1989.

Drake, Stillman. *Galileo Studies: Personality, Tradition, and Revolution*. Ann Arbor: University of Michigan Press, 1970.

Dybas, Cheryl. "Light from the Shadows of Distant Proto-Galaxies Discovered by NSF Astronomers." Press Release. Washington, DC: National Science Foundation, January 7, 1993.

Eddington, Arthur S. *The Nature of the Physical World*. London: Cambridge University Press, 1932.

Eicher, David J. "Keck's First Light." *Astronomy* (April 1991): 42–44.

Eisler, Riane. *The Chalice and the Blade: Our History, Our Future*. San Francisco: Harper & Row, 1987.

Elvee, Richard Q., ed. *Mind in Nature*. New York: Harper and Row, 1982.

Feinberg, Gerald. *What Is the World Made Of? Atoms, Leptons, Quarks and Other Tantalizing Particles*. New York: Doubleday, 1978.

Feynman, Richard. *The Feynman Lectures on Physics*. 3 vols. Reading, MA: Addison-Wesley, 1963–1965.

———. *QED: The Strange Theory of Light and Matter*. Princeton, NJ: Princeton University Press, 1985.

Feynman, Richard, with Ralph Leighton. *"Surely You're Joking, Mr. Feynman": Adventures of a Curious Character*. New York: W.W. Norton, 1985.

———. *What Do You Care What Other People Think? Further Adventures of a Curious Character*. New York: W.W. Norton, 1988.

Field, George B., and Eric J. Chaisson. *The Invisible Universe: Probing the Frontiers of Astrophysics*. New York: Vintage Books, 1985.

Flam, Faye. "COBE Finds the Bumps in the Big Bang." *Science* (May 1, 1992): 612.

———. "COBE Sows Cosmological Confusion." *Science* (July 3, 1992): 28–30.

———. "A Galaxy Is Born." *Science* (1991): 1294.

———. "Listening to the Music of the Spheres." *Science* (September 13, 1991): 1207–1208.

———. "Looking Toward the Edge." *Science* (July 12, 1991): 139.

———. "Mirror, Mirror, Which Is the Fairest?" *Science* (April 23, 1993): 493–494.

Folger, Tim. "The Big Eye." *Discover* (November 1991): 40–44.

Forward, Robert L., and Joel Davis. *Mirror Matter: Pioneering Antimatter Physics*. New York: Wiley, 1988.

Gamow, George. *Matter, Earth, and Sky*. 2nd ed. Englewood Cliffs, NJ: Prentice-Hall, 1965.

———. *My World Line: An Informal Autobiography*. New York: Viking Press, 1970.

———. *Thirty Years That Shook Physics*. New York: Doubleday, 1966.

Gardner, Martin. "Illusions of the Third Dimension." *Psychology Today* (August 1983): 62–67.

Gifford, Don. *The Farther Shore: A Natural History of Perception, 1798–1984*. New York: Atlantic Monthly Press, 1990.

Gimbutas, Marija. *The Goddesses and Gods of Old Europe*. Berkeley, CA: University of California Press, 1982.

———. *The Language of the Goddess*. San Francisco: HarperSanFrancisco, 1989.

Gleick, James. *Genius: The Life and Science of Richard Feynman*. New York: Pantheon Books, 1992.

Goldstein, Thomas. *Dawn of Modern Science*. Boston: Houghton Mifflin, 1980.

Graham, David. "A Sharper Image of the Cosmos." *Technology Review* (October 1992): 12–13.

Gregory, Bruce. *Inventing Reality: Physics as Language*. New York: Wiley, 1988.

Gregory, Richard L. *Eye and Brain: The Psychology of Seeing*. 4th ed. Princeton, NJ: Princeton University Press, 1990.

———, ed. *The Oxford Companion to the Mind*. New York: Oxford University Press, 1987.

Gribbin, John. "Bunched Red Shifts Question Cosmology." *New Scientist* (December 21, 1991): 10.

———. *In Search of the Big Bang: Quantum Physics and Cosmology*. New York: Bantam Books, 1986.

Haldane, J. B. S. *Possible Worlds and Other Papers*. New York: Harper & Brothers, 1928.

Halliwell, Jonathan J. "Quantum Cosmology and the Creation of the Universe." *Scientific American* (December 1991): 76–85.

Handingham, Evan. *Early Man and the Cosmos*. New York: Walker Publishing, 1984.

Harris, Errol E. *Cosmos and Anthropos: A Philosophical Interpretation of the Anthropic Cosmological Principle*. London: Humanities Press International, 1991.

Harris, Joel K. "An Optical Revolution in Chile." *Astronomy* (November 1990): 44–49.

———. "Seeing a Brave New World." *Astronomy* (August 1992): 22–29.

Harrison, Edward. "Newton and the Infinite Universe." *Physics Today* (February 1986): 24–30.

Hawkins, Gerald. *Beyond Stonehenge*. New York: Harper & Row, 1973.

———. *Stonehenge Decoded*. New York: Doubleday, 1965.

Heisenberg, Werner. *Physics and Beyond*. New York: Harper & Row, 1971.

———. *Physics and Philosophy: The Revolution in Modern Science*. New York: Harper & Row, 1958.

Hellemans, Alexander, and Bryan Bunch. *The Timetables of Science: A Chronology of the Most Important People and Events in the History of Science*. New York: Simon and Schuster, 1988.

Herbert, Nick. *Quantum Reality*. New York: Doubleday, 1985.

Hoffman, Banesh. *The Strange Story of the Quantum*. New York: Dover, 1959.

Horgan, John. "COBE Corroborated." *Scientific American* (February 1993): 22.

———. "The End of Clumpiness?" *Scientific American* (February 1992): 26.

———. "Eternally Self-Reproducing Universe?" *Scientific American* (April 1993): 24.

———. "Questioning the 'It' From 'Bit.'" *Scientific American* (June 1991): 36–38.

Hovis, R. Corby, and Helge Kragh. "P.A.M. Dirac and the Beauty of Physics." *Scientific American* (May 1993): 104–109.

Hubel, David H. *Eye, Brain, and Vision*. New York: W.H. Freeman, 1988.

"Infrared Image Reveals Possible Brown Dwarf." *Sky & Telescope* (July 1988): 12.

Jaki, Stanley L. *Cosmos in Transition: Studies in the History of Cosmology*. Tucson, AZ: Pachart Publishing House, 1990.

Jaroff, Leon. "Roaming the Cosmos." *Time* (March 1992): 250–251.

"Keck Foundation Funds Duplicate of Keck Telescope." *Physics Today* (July 1991): 52–53.

Klauder, John R., ed. *Magic without Magic: John Archibald Wheeler*. San Francisco: W.H. Freeman, 1972.

Kneller, George F. *Science as a Human Endeavor*. New York: Columbia University Press, 1978.

Krauss, Lawrence M. "Dark Matter in the Universe." *Scientific American* (December 1986): 58–68.

Krupp, E. C., ed. *In Search of Ancient Astronomies*. New York: Doubleday, 1977.

Lake, George. "Cosmology of the Local Group." *Sky & Telescope* (December 1992): 613–619.

LeShan, Lawrence, and Henry Margenau. *Einstein's Space and Van Gogh's Sky: Physical Reality and Beyond*. New York: Macmillan, 1982.

Learner, Richard. *Astronomy through the Telescope*. New York: Van Nostrand Reinhold, 1981.

Leibovic, K. N., ed. *Science of Vision*. New York: Springer Verlag, 1990.

Levi, Barbara Goss. "New Technology Telescope Actively Corrects for Misalignments." *Physics Today* (May 1990): 17–18.

Levine, Michael W., and Jeremy M. Shefner. *Fundamentals of Sensation and Perception*. 2nd ed. Pacific Grove, CA: Brooks/Cole, 1991.

Lightman, Alan. *Time for the Stars: Astronomy in the 1990s*. New York: Viking Press, 1992.

Magrath, Barney. "Optical Astronomy Looks to the Future." *Astronomy* (November 1990): 34–32.

Martin, Buddy, et al. "The New Ground-Based Optical Telescopes." *Physics Today* (March 1991): 22–30.

Mather, John C., and John Boslough. *The Very First Light: The True Inside Story of the Scientific Journey Back to the Dawn of the Universe*. New York: Basic Books, 1996.

Microsoft Bookshelf '95. Redmond, WA: Microsoft Corp., 1995.

Misner, Charles W., Kip S. Thorne, and John A. Wheeler. *Gravitation*. San Francisco: W.H. Freeman, 1973.

Moore, Patrick. *Patrick Moore's History of Astronomy*. 6th rev. ed. London: Macdonald, 1983.

Morris, Richard. *The Edges of Science: Crossing the Boundary from Physics to Metaphysics*. New York: Prentice-Hall, 1990.

Nelson, Jerry. "The Keck Telescope." *American Scientist* (March 1989): 170–176.

"New U.S. Telescope for the Southern Hemisphere." *Sky & Telescope* (October 1991): 344.

Noyes, Russell. *William Wordsworth*. New York: Twayne Publishers, 1971.

Nye, Mary Jo, Joan Richards, and Roger Stuewer, eds. *The Invention of Physical Science: Intersections of Mathematics, Theology and Natural Philosophy since the Seventeenth Century*. Boston: Kluwer Academic, 1992.

Ostling, Richard K. "Galileo and Other Faithful Scientists." *Time* (December 2, 1992): 42–43.

Paananen, Victor N. *William Blake*. New York: Twayne Publishers, 1977.

Pais, Abraham. *Inward Bound: Of Matter and Forces in the Physical World*. New York: Oxford University Press, 1986.

Pannekoek, A. *A History of Astronomy*. New York: Dover, 1961, 1989.

Poetical Works of John Keats. Home Page. http://www.cc.columbia. edu/ acis/ bartleby/keats. New York: Columbia University, 1996.

Pool, M. D., and J. Lawrence. *Nature's Masterpiece: The Brain and How It Works*. New York: Walker, 1987.

Polkinghorne, J. C. *The Quantum World*. Princeton, NJ: Princeton University Press, 1984.
Powell, Corey S. "The Golden Age of Cosmology." *Scientific American* (July 1992): 17–22.
———. "Inconstant Cosmos." *Scientific American* 268 (May 1993): 110–118.
———. "MACHOs or WIMPs?" *Scientific American* (January 1993): 27–29.
———. "Mirroring the Cosmos." *Scientific American* (November 1991): 80–89.
Purkis, John. *A Prelude to Wordsworth*. New York: Charles Scribner's Sons, 1970.
"Quantum Gravity." *Scientific American* (December 1983): 112–129.
"The Quantum Universe: A Zero-Point Fluctuation?" *Science News* (August 3, 1985): 72–74.
"The Quantum Wave Function of the Universe." *Science* (December 2, 1988): 1248–1250.
Quennell, Peter. *Romantic England: Writing and Painting 1717–1851*. New York: Macmillan, 1970.
Rae, Alastair. *Quantum Physics: Illusion or Reality*? New York: Cambridge University Press, 1986.
Reed, Mark A. "Quantum Dots." *Scientific American* (January 1993): 118–123.
"Religion and the Big Bang Discovery." *Seattle Times* (April 25, 1992): B6.
Renteln, Paul. "Quantum Gravity." *American Scientist* (November 1991): 508–527.
Ressmeyer, Roger H. "Keck's Giant Eye." *Sky & Telescope* (December 1992): 623–25.
Ronan, Colin. *The Atlas of Scientific Discovery*. London: Quill Publishing, 1983.
———. *The Natural History of the Universe: From the Big Bang to the End of Time*. New York: Macmillan, 1991.
Ruthen, Russell. "The Cosmic Microwave Mirage?" *Scientific American* (October 1992): 27.
Seigfried, Tom. "Satellite's Look Backward Hurls Science Forward." *Seattle Times* (May 4, 1992): F1, F5.
Selected Poetry of George Gordon, Lord Byron (1788–1824). Home Page. http://library.utoronto.ca/www/utel/rp/poems/byron10.html. Toronto: University of Toronto, Department of English, 1996.
Silk, Joseph. *The Big Bang*. New York: W.H. Freeman, 1989.
Singer, Charles. *From Magic to Science: Essays on the Scientific Twilight*. New York: Dover, 1958.
Sinnott, Roger W. "The Keck Telescope's Giant Eye." *Sky & Telescope* (July 1990): 15–22.
Small, Christopher. *Mary Shelley's "Frankenstein": Tracing the Myth*. Pittsburgh: University of Pittsburgh Press, 1973.

"Smithsonian to Build Unique Telescope on Mauna Kea." Press Release. Cambridge, MA: Harvard College Observatory, May 27, 1991.

Some Poems from Complete Poetical Works (1888). *Home Page. http://www.cc.columbia.edu/acis/bartleby/wordsworth/ww138.html.* New York: Columbia University, 1996.

"Special Report: Optics Clear Air for Astronomers." *Popular Mechanics* (November 1989): 16.

Starhawk. *The Spiral Dance.* New York: Harper & Row, 1979.

Starhawk. *Truth or Dare: Encounters with Power, Authority, and Mystery.* San Francisco: HarperSanFrancisco, 1987.

Sternheim, Morton M., and Joseph W. Kane. *General Physics.* New York: Wiley, 1986.

Stewart, R. J. *The Elements of Creation Myth.* Longmead, England: Element Books, 1989.

Stone, James. "New Window on the Cosmos." *Sky & Telescope* (July 1990): 42–43.

Stryer, Lubert. "The Molecules of Visual Extinction." *Scientific American* (July 1987): 42–50.

Swimme, Brian. *The Universe Is a Green Dragon.* Santa Fe, NM: Bear & Company, 1984.

Talcott, Richard. "COBE's Big Bang!" *Astronomy* (August 1992): 42–44.

Tauber, Gerald E. *Man's View of the Universe: A Pictorial History.* New York: Crown Publishers, 1979.

Taubes, Gary. "OGLEing, MACHOs, and the Search for Dark Matter." *Science* (April 23, 1993): 492–93.

Thorne, Kip, and Wokciech H. Zureck. "John Archibald Wheeler: A Few Highlights of His Contributions to Physics." In *Between Quantum and Cosmos: Studies and Essays in Honor of John Archibald Wheeler*, edited by Wojciech Hubert Zurek et al. Princeton, NJ: Princeton University Press, 1988.

Tiefert, Marj, ed. *Hypertext Poems from the Coleridge Archive. Home Page. http://www.lib.virginia.edu/etext/stc/Coleridge/poems.* 1996.

Travis, John. "The Hubble Constant Takes the Low Road Again." *Science* (July 3, 1992): 34.

Trefil, James S. *Space, Time, Infinity: The Smithsonian Views the Universe.* New York: Pantheon Books, 1985.

Unguru, Sabetai, ed. *Physics, Cosmology and Astronomy, 1300–1700: Tension and Accommodation.* Boston: Kluwer Academic, 1991.

Wald, George. "Life and Light" *Scientific American* (October 1959): 92–1081.

Waldrop, M. Mitchell. "Astronomers Try to Put Mauna Kea into Space." *Science* (August 31, 1990): 987.

———. "Computer-Age Stargazing." *Science* (September 15, 1989): 1191.

———. "Keck's First Light." *Science* (December 14, 1990): 1511.

———. "The New Art of Telescope Making." *Science* (December 19, 1986): 1495–1497.

Weaver, Jefferson Hane, ed. *The World of Physics: A Small Library of the Literature of Physics from Antiquity to the Present*. vol. 1. New York: Simon and Schuster, 1987.

Wheeler, John Archibald. "Bohr, Einstein, and the Strange Lesson of the Quantum." In *Mind in Nature*, edited by Richard Q. Elvee. New York: Harper and Row, 1982.

———. "The Computer and the Universe." *International Journal of Theoretical Physics* 21 (1982).

———. *The Frontiers of Time*. Amsterdam: North-Holland, 1979.

———. "Law without Law." In *Quantum Theory and Measurement*, edited by John A. Wheeler and W. H. Zurek. Princeton, NJ: Princeton University Press, 1983.

———. "On the Nature of Quantum Geometrodynamics." *Annals of Physics* (1957): 604–614.

———. "Our Universe: The Known and the Unknown." *American Scientist* (1968): 1–13.

———. "The Quantum and the Universe." In *Relativity, Quanta, and Cosmology*. Vol. 2, edited by M. Pantaleo and F. de Finis. New York: Johnson Reprint Co., 1979.

Wheeler, John Archibald, and W. H. Zurek, eds. *Quantum Theory and Measurement*. Princeton, NJ: Princeton University Press, 1983.

Wigner, Eugene P. "Remarks on the Mind-Body Question." In *Quantum Theory and Measurement*, edited by John A. Wheeler and W. H. Zurek. Princeton, NJ: Princeton University Press, 1983.

———. "The Limitations of the Validity of Present-Day Physics." In *Mind in Nature*, edited by Richard Q. Elvee. New York: Harper and Row, 1982.

Wilford, John Noble. "Scientists Report Profound Insight on How Time Began." *New York Times* (April 24, 1992): A1, A11.

Wolf, Fred Alan. *Taking the Quantum Leap*. San Francisco: Harper & Row, 1981.

Woolf, Harry, ed. *Some Strangeness in the Proportion: A Centennial Symposium to Celebrate the Achievements of Albert Einstein*. Reading, MA: Addison-Wesley, 1980.

Wright, Robert. "Science, God and Man." *Time* (December 28, 1992): 38–44.

Zukav, Gary. *The Dancing Wu Li Masters*. New York: Bantam Books, 1980.

Zurek, Wojciech Hubert, et al., eds. *Between Quantum and Cosmos: Studies and Essays in Honor of John Archibald Wheeler*. Princeton, NJ: Princeton University Press, 1988.

Index